T0396869

Analytical Fluid Dynamics in Lagrangian Variables

Analytical Fluid Dynamics in Lagrangian Variables

Abrashkin Anatoly Alexandrovich
National Research University Higher School of Economics,
Nizhny Novgorod, Russia

NEW JERSEY • LONDON • SINGAPORE • GENEVA • BEIJING • SHANGHAI • TAIPEI • CHENNAI

Published by

World Scientific Publishing Co. Pte. Ltd.
5 Toh Tuck Link, Singapore 596224
USA office: 27 Warren Street, Suite 401-402, Hackensack, NJ 07601
UK office: 57 Shelton Street, Covent Garden, London WC2H 9HE

Library of Congress Control Number: 2024056900

British Library Cataloguing-in-Publication Data
A catalogue record for this book is available from the British Library.

ANALYTICAL FLUID DYNAMICS IN LAGRANGIAN VARIABLES

Copyright © 2025 by World Scientific Publishing Co. Pte. Ltd.

All rights reserved. This book, or parts thereof, may not be reproduced in any form or by any means, electronic or mechanical, including photocopying, recording or any information storage and retrieval system now known or to be invented, without written permission from the publisher.

For photocopying of material in this volume, please pay a copying fee through the Copyright Clearance Center, Inc., 222 Rosewood Drive, Danvers, MA 01923, USA. In this case permission to photocopy is not required from the publisher.

ISBN 978-981-98-0633-1 (hardcover)
ISBN 978-981-98-0634-8 (ebook for institutions)
ISBN 978-981-98-0635-5 (ebook for individuals)

For any available supplementary material, please visit
https://www.worldscientific.com/worldscibooks/10.1142/14133#t=suppl

Desk Editors: Kannan Krishnan/Rhaimie Wahap

Typeset by Stallion Press
Email: enquiries@stallionpress.com

Preface

This book is devoted to the problems of the analytical description of incompressible fluid flows in Lagrangian variables. Commonly, specialists in fluid dynamics prefer to use the Euler approach. For example, even classical textbooks by J. Batchelor, L.G. Loitsyansky, L.D. Landau, and E.M. Lifshitz do not provide the Lagrange equations of fluid dynamics. Exceptions are the fundamental work by G. Lamb and the two-volume work by N.E. Kochin, I.A. Kibel, and N.V. Rose, but even there, only a few paragraphs are devoted to Lagrangian variables.

An extreme point of view on this issue was expressed by D. Küchemann in his book *The Aerodynamic Design of Aircraft*. In his opinion, using the Lagrangian description in practice is inconvenient and there is usually no need for it. This observation needs to be commented upon. D. Küchemann was engaged in aircraft design. Indeed, it is inconvenient to use Lagrangian variables to solve flow problems, but they turn out to be preferable to Eulerian variables in some cases when other types of flows are described. This applies, in particular, to the problems of vortex dynamics and flows with a free boundary.

I am inclined to consider the Lagrange method as an additional (to the Euler method) technique for studying fluid motion. It is well known that for most exact solutions, the transition from one description method to another is simply impossible, and each method is a different mathematical reflection of the real flow. Moreover, the natural (deep) essence of fluid dynamics is related to tracing the trajectories of the motion of specific fluid particles. This type of analysis is exactly what the Lagrange method offers.

The purpose of this publication is to reflect as fully as possible on the achievements of analytical fluid dynamics in Lagrangian variables. A significant part of it consists of my original articles. Considerable attention has been paid to the intricacies of the Lagrangian description and its popularization among fluid dynamics specialists, undergraduates, graduate students, and anyone interested in problems of fluid mechanics. This book can be considered, among other things, a tutorial on Lagrangian fluid dynamics.

In 2006, Cambridge University Press published A. Bennett's book *Lagrangian Fluid Dynamics*. This was the first book published in Western countries devoted exclusively to the Lagrangian description of fluid motion. Although the title is similar, it differs from the present book in both scope and presentation. While Bennett's book devotes considerable attention to the statistical description of fluids (more than half the book), I focus exclusively on problems of classical fluid dynamics. This publication can be a useful addition to well-known fluid dynamics courses that use only Eulerian variables.

I would like to express my gratitude to Anastasiya Volodina and Alexey Krayev for translating this book.

About the Author

Abrashkin Anatoly Alexandrovich is a graduate of Gorky State University (1981), Doctor of Philosophy (1987), and Doctor of Physical and Mathematical Sciences (2000). He is Professor at the Department of Mathematics, National Research University Higher School of Economics (Nizhny Novgorod, Russia) and a specialist in theoretical fluid dynamics. He is the author of more than 60 scientific articles in peer-reviewed journals and has also coauthored the monograph "Vortex dynamics in Lagrangian description" (2006, in Russian) with E.I. Yakubovich.

Contents

Preface   v

About the Author   vii

Part I. Description of Basics   1

Chapter 1.   Ideal Fluid Equations   3

1.1.   Eulerian and Lagrangian Coordinates   3
1.2.   Continuity Equation   7
1.3.   Momentum Equations   9
     1.3.1.   Vector and coordinate forms   9
     1.3.2.   Weber momentum equations   11
     1.3.3.   Cauchy momentum equations   12
     1.3.4.   Matrix form of momentum equations   14
1.4.   Vorticity and Cauchy Invariants   15
1.5.   Physical Meaning of Cauchy Invariants   18
1.6.   Fluid Dynamics Equations in Curvilinear Coordinate
     Systems   21
     1.6.1.   Cylindrical geometry   22
     1.6.2.   Spherical geometry   23
References   23

Chapter 2.   Potential Motion   27

2.1.   Flow Potentiality Conditions   27
2.2.   Flow Near the Critical Point   28
     2.2.1.   Two-dimensional motion   29
     2.2.2.   Axisymmetric case   30

Analytical Fluid Dynamics in Lagrangian Variables

2.3.	Flow Initiated by a Point Vortex	32
2.4.	Fluid Ellipse Dynamics	33
2.5.	Movement of a Fluid Cone with a Free Side Surface	38
2.6.	Spherical Bubble Dynamics	41
2.7.	John's Method	43
	2.7.1. Exact solution for a vortex-free wave	45
	2.7.2. John's parabola	48
	2.7.3. Ascending flow	49
References		51

Chapter 3.   Vortex Motion   53

3.1.	Complex form of Fluid Dynamics Equations	53
3.2.	Jacobian form of Equations for Two-Dimensional Flows	54
3.3.	The Simplest Types of Vortex Flows	55
	3.3.1. Inertial flows	55
	3.3.2. Flows with circular streamlines	56
	3.3.3. Flow inside the ellipse	57
	3.3.4. Flows with helical particle trajectories	58
3.4.	The Kirchhoff Vortex	59
3.5.	Unsteady Motion of a Flat Layer with a Free Boundary	61
3.6.	Spatial Flows with a Velocity Field Linearly Dependent on Coordinates	62
	3.6.1. Dirichlet problem	63
	3.6.2. Ovsyannikov ellipsoid	64
References		68

Part II. Waves on Water   69

Chapter 4.   Potential Approximation   71

4.1.	Stokes Wave	72
	4.1.1. The method of modified Lagrangian coordinates	72
	4.1.2. Linear wave	75
	4.1.3. Stokes drift: Nonlinear dispersion relation	77
	4.1.4. Overview of other works	78
4.2.	Wave Train of Surface Gravity Waves	80

Contents xi

4.3. Dam Failure Problem 84
References 89

Chapter 5. Gerstner Wave 91

5.1. Basic Properties 92
    5.1.1. Implementation problem 95
5.2. Edge Waves on a Sloping Beach 97
5.3. Waves in Layers with Density Discontinuities 99
References 102

Chapter 6. Weakly Vortex Waves 107

6.1. Guyon Waves on Deep Water 107
    6.1.1. Linear approximation 109
    6.1.2. Quadratic approximation 114
    6.1.3. Cubic approximation 118
6.2. Development of the Modified Coordinates Method 118
    6.2.1. Stationary spatial motion 118
    6.2.2. Non-stationary flow 120
6.3. Nonlinear Schrödinger Equation for Quadratic
    Vortex Waves 120
References 122

Part III. Exact Solutions in Vortex Dynamics 125

Chapter 7. Two-Dimensional Vortex Flow 127

7.1. Review of Known Approaches in the Study
    of Plane Vortex Flows 127
7.2. Ptolemaic Fluid Motion Class 130
7.3. The Uniqueness of Ptolemaic Flows 132
7.4. Gerstner Waves on a Cylindrical Surface 136
References 139

Chapter 8. Localized Two-Dimensional
Vortices in the External Flow 143

8.1. Ptolemaic Vortices 145
8.2. Nonlinear Kelvin Vortex Waves 148
    8.2.1. Problem statement and solution method 149
    8.2.2. Examples 151

xii *Analytical Fluid Dynamics in Lagrangian Variables*

8.3. Interaction of Two Elliptical Vortices 152
    8.3.1. Generalized Kirchhoff vortices 154
    8.3.2. Time-harmonic deformation field 157
8.4. Non-Lagrangian Approaches 163
References 164

Chapter 9.   Spatial Vortex Flows   167

9.1. Complex form of Matrix Equations of Fluid
    Dynamics 168
9.2. Generalized Ptolemaic Flows 170
    9.2.1. Flows with straight vortex lines 171
    9.2.2. Flows with vortex lines in the same plane
         with the rotation axis 174
    9.2.3. Flows for which the rotation axis is inclined
         to the vortex line plane 178
9.3. Ptolemaic Vortices in the Axial Flow 182
References 184

Part IV. Generalized Gerstner Waves   187

Chapter 10.   Waves on Water at Unstable Pressure   189

10.1. Oscillating Standing Soliton 191
10.2. Soliton Against the Background of Uniform
     Undulation 194
10.3. Non-Stationary Gerstner Waves 196
References 202

Chapter 11.   Vortex Model of Rogue Waves   205

11.1. Birth from the Gerstner Wave 206
11.2. Special Pressure Mode 210
    11.2.1. Stationarity of pressure in the Ptolemaic
         flow 211
    11.2.2. Quasi-stationary pressure pit 214
References 216

Chapter 12.   Breaking of the Surface Gravity Wave   219

12.1. Deep-Water Breather Dynamics 220

Contents

xiii

12.2. The Breather Breaking Mechanism 225
References 228

Chapter 13. Non-Stationary Edge Waves 229

13.1. 3D Exact Solution 232
13.2. Gerstner–Constantin Solution 235
13.3. Gerstner-Like Waves 237
References 243

Part V. Viscous Fluid Flows 245

Chapter 14. Basic Equations and Examples 247

14.1. Navier–Stokes Equations 247
14.2. Equations for Cauchy Invariants 250
14.3. Exact Solutions 253
    14.3.1. Flow in the channel and the Couette flow 253
    14.3.2. Hagen–Poiseuille flow in a round tube 255
    14.3.3. A flat wall suddenly set in motion
        (Stokes' first problem) 256
    14.3.4. Flow created by a vibrating plane
        (Stokes' second problem) 258
    14.3.5. Flow between two rotating cylinders 259
    14.3.6. Vortex diffusion 261
    14.3.7. Diffuser flow 261
References 262

Chapter 15. Gravity Waves on the Surface of a Viscous Fluid 265

15.1. Standing Waves 266
    15.1.1. Linear spatial oscillations 266
    15.1.2. Long-wave approximation 270
    15.1.3. Nonlinear two-dimensional waves 272
15.2. Traveling Spatial Waves 275
    15.2.1. Navier–Stokes equations in modified
        Lagrangian coordinates 275
    15.2.2. Averaged drift 277
15.3. A Train of Two-Dimensional Gravity Surface Waves
    for Large Reynolds Numbers 279
References 283

xiv   *Analytical Fluid Dynamics in Lagrangian Variables*

Part VI. The Earth's Rotation Effect   285

Chapter 16. Exact Solutions for Waves   287

16.1. Gerstner Waves in a Rotating Fluid 289
    16.1.1. Pollard's solution 289
    16.1.2. Nonlinear drift flows generated by wind
        in Arctic regions 292
16.2. Waves in the Equatorial Region 296
    16.2.1. f-plane approximation 297
    16.2.2. Trapped waves (β-plane approximation) 299
    16.2.3. Accounting for stratification 301
References 302

Chapter 17. Vortices in a Rotating Fluid   305

17.1. Motion of Unsteady Columns in an Exponentially
    Stratified Fluid in the Boussinesq Approximation 305
17.2. Lagrangian Equations in Spherical Coordinates 307
17.3. Problem Statement and Approximations Used 312
    17.3.1. Thin layer approximation 312
    17.3.2. Averaged latitude approximation 314
    17.3.3. Ptolemaic vortices on the sphere 317
References 323

*Appendix 1. Speed and Acceleration in a Cylindrical
    Coordinate System*   325

*Appendix 2. Speed and Acceleration in a Spherical
    Coordinate System*   327

Index   329

Part I
Description of Basics

Chapter 1

Ideal Fluid Equations

1.1. Eulerian and Lagrangian Coordinates

> "It is only to Euler that we owe the first
> general formulas for the motion of fluids..."
>
> — Joseph Louis Lagrange. Analytical mechanics

The equations of fluid momentum can be represented in two different forms, depending on whether we are interested in the values of the flow parameters at an arbitrary point in space or are trying to determine the "history" of an individual particle. The first form was proposed by Leonhard Euler in his memoirs "General Principles of Fluid Motion" [1] and is usually called Eulerian. The equations of fluid dynamics for an ideal fluid in the Eulerian formulation have the form [2]

$$\frac{\partial \rho}{\partial t} + \operatorname{div} \rho \vec{V} = 0; \tag{1.1}$$

$$\frac{\partial \vec{V}}{\partial t} + (\vec{V}\nabla)\vec{V} = -\frac{1}{\rho}\nabla p + \vec{f}. \tag{1.2}$$

Here, $\vec{V}(x, y, z, t)$ is the velocity field at points with Cartesian coordinates x, y, and z at time t, ρ is the fluid density, p is the pressure, and $\vec{f}$ is an external force acting on a unit mass. The first of these relations is called the continuity equation and the second is the Euler equation or the momentum equation.

(a) (b)

Fig. 1.1. (a) Leonhard Euler (1707–1783) and (b) Joseph Louis Lagrange (1736–1813), the founders of the Lagrangian approach in fluid dynamics.

Euler should also be considered the author of the second form, based on the observation of the motion of fluid particles which is attributed to Lagrange (see Fig. 1.1). In a letter to Lagrange dated January 1, 1760, Euler pointed out the possibility of choosing fluid dynamics variables related to the coordinates of fluid particles. In 1762, Lagrange published Euler's aforementioned letter, which personally confirmed his discovery of a second approach to describing fluid motion. The name "Lagrangian" for it was proposed by Dirichlet in 1860 [3, 4]. This "erroneous terminology" [3] has become established in modern fluid dynamics.

The Lagrangian way of describing the motion of a fluid is based on the motion of fixed particles of the fluid, which can be traced starting from a certain point in time. While the Eulerian approach studies the velocity field at a point with x, y, and z coordinates (Eulerian variables), the Lagrangian approach traces each fluid particle by first labeling ("numbering") it. Thus, the coordinates of the fluid particles will be unknown functions, and we will denote them as X, Y, and Z, and the "numbers" of particles and time will be the independent variables. The choice of these labels is largely arbitrary; it is only necessary to ensure that the numbering is one to one and that all particles are "counted." For example, these may be the initial coordinates X_0, Y_0 and Z_0 of fluid particles, but this is only one of the options for introducing Lagrangian variables.

Let us give their general definition. We denote the Cartesian components of the flow velocity field $u(x, y, z, t), v(x, y, z, t)$, and $w(x, y, z, t)$. Then, the following equalities will be valid:

$$dt = \frac{dx}{u} = \frac{dy}{v} = \frac{dz}{w}. \tag{1.3}$$

Here, x, y, and z are already considered as functions of time t. Obviously, this is the essence of the equation from which particle trajectories are determined. The solutions of Eqs. (1.1) and (1.2) can be represented in the form

$$x = X(a, b, c, t), \quad y = Y(a, b, c, t), \quad z = Z(a, b, c, t), \tag{1.4}$$

where a, b, and c are some constants that determine the position of fluid particles. In a particular case, they can be chosen to coincide with the coordinates of fluid particles at some initial time t_0, so that

$$a = X|_{t=t_0}, \quad b = Y|_{t=t_0}, \quad c = Z|_{t=t_0}. \tag{1.5}$$

However, in the most general case, each variable a, b, and c is some function of the coordinates of the initial position of the fluid particle. Three constants in Eq. (1.4) are used as labels for an individual fluid particle. We will call them Lagrangian coordinates.

Truesdell opposed such terminology in 1954. He proposed, based on a letter to Euler, to call the coordinates x, y, and z spatial and the labels of particles a, b, and c material [3]. Such innovations did not take root. However, the term "labeled" is intensely used in the papers by American scientists in relation to Lagrangian variables. In our opinion, the fight against established terminology and the actual belittlement of Lagrange's contribution to the creation of one of the approaches to fluid dynamics are hardly justified. This point of view received worldwide recognition thanks to his "Analytical Mechanics" [5], published in 1788, and Dirichlet had all the ethical grounds to assign the name "Lagrangian" to material variables.

In the Eulerian space of variables x, y, and z, the lines corresponding to the axes of the Lagrangian coordinate system will, generally speaking, be complex non-stationary curves. Consider, e.g., a set of particles with coordinates $b = 0$ and $c = 0$, i.e., located along the a coordinate variation line. Over time, this fluid "axis" will bend and move in space. The same can be said about the b and c "axes." They can be considered the axes of a "fluid" coordinate system. This system is generally curved and non-orthogonal; in a moving fluid, its axes are continuously deformed, following its constituent particles. In other words, this system is "frozen" into a fluid medium. Therefore, the essence of the Lagrangian description in fluid dynamics is the introduction of a coordinate system frozen into the fluid.

"The Lagrange method is most directly related to the real movement of individual fluid elements, the totality of which constitutes the flow; therefore, it can be considered physically more *natural* (author's italics — A.A.) than the Euler description method." [6] However, Lagrangian variables are rarely used in fluid mechanics. This is due to both the complex form of nonlinearity of the Lagrange fluid dynamics equations and their bulkiness (especially in the case of a viscous fluid). Unlike the Eulerian approach, in Lagrangian variables, both the momentum equations and the continuity equation have a nonlinear form.

However, the advantages of using Lagrangian coordinates are often discovered at the level of problem formulation. A significant advantage of the Lagrange equations is that their solution should be sought in a given domain of the space of variables. In particular, when considering waves on water, it can be assumed that the vertical Lagrangian coordinate value equal to zero corresponds to the unknown shape of the free surface and the lower half-space of Lagrangian variables to the entire region of wave motion. Another well-known fact is that in the plane flow of an ideal incompressible fluid, the vorticity of the fluid particles is preserved. As a result, when studying the dynamics of vortex regions (localized vortices, waves on water, etc.), it is more convenient to specify the initial vorticity distribution precisely in Lagrangian coordinates (for example, inside a unit circle). In the following, these provisions will be discussed in more detail when considering specific flows.

To turn from Lagrangian to Eulerian variables, it is necessary to differentiate the functions X, Y, and Z in time, thereby obtaining the expressions for the velocity components as functions of Lagrangian variables,

$$u = X_t(a, b, c, t); \quad v = Y_t(a, b, c, t); \quad w = Z_t(a, b, c, t), \qquad (1.6)$$

Ideal Fluid Equations

and then solving the system of Eq. (1.4) for these variables, i.e., finding the dependences

$$a = a(x, y, z, t); \quad b = b(x, y, z, t); \quad c = c(x, y, z, t),$$

and substituting them into Eq. (1.6). It is relatively easy to implement such a program for weakly nonlinear flows, but this happens extremely rarely for exact solutions of fluid dynamics. It can be said that the Eulerian and Lagrangian approaches exist as two complementary ways of describing fluid motion.

1.2. Continuity Equation

We assume that a moving fluid fills space completely or a certain part of it and that during movement there is neither loss of matter nor its appearance (there are no sources or sinks). This assumption imposes a certain condition on the change in density and volume of the fluid, called the continuity equation.

Consider two positions of the same volume of fluid at times t_0 and t. Let the fluid volume τ_0 at the moment of time t_0 turns into the volume τ at the moment of time t. A fluid particle with coordinates

$$x_0 = X(a, b, c, t_0); \quad y_0 = Y(a, b, c, t_0); \quad z_0 = Z(a, b, c, t_0), \quad (1.7)$$

at time t_0 will shift at time t to the position with coordinates x, y, and z given by Eq. (1.4). The fluid density at the initial moment $\rho_0(a, b, c, t_0)$ at time t will be $\rho(a, b, c, t)$. However, the volumetric mass of the fluid does not change during its movement; therefore, the following relation is valid:

$$\iiint_{\tau_0} \rho_0 dx_0 dy_0 dz_0 = \iiint_{\tau} \rho dx dy dz. \quad (1.8)$$

Let us proceed to integration over Lagrangian variables in both integrals. According to the well-known rules for transforming an integral

8   *Analytical Fluid Dynamics in Lagrangian Variables*

when changing variables, we have

$$\iiint_{\tau_0} \rho_0 dx_0 dy_0 dz_0 = \iiint_{\tau_0} \rho_0 D_0 da\,db\,dc;$$

$$\iiint_{\tau} \rho\,dx\,dy\,dz = \iiint_{\tau_0} \rho D\,da\,db\,dc,$$

(1.9)

where D_0, and D are the Jacobians of the transformations

$$D_0 = \frac{D(x_0, y_0, z_0)}{D(a, b, c)}; \quad D = \frac{D(x, y, z)}{D(a, b, c)} = \frac{D(X, Y, Z)}{D(a, b, c)} = \begin{vmatrix} X_a & X_b & X_c \\ Y_a & Y_b & Y_c \\ Z_a & Z_b & Z_c \end{vmatrix};$$

hereafter, the subscripts in the determinant mean taking the derivative with respect to the corresponding variable.

Moving all terms onto one side of Eq. (1.8) and applying Eq. (1.10), we find

$$\iiint_{\tau_0} (\rho_0 D_0 - \rho D)\,da\,db\,dc = 0.$$

Hence, due to the arbitrariness of the initially taken volume τ_0, at any given time, a relation

$$\rho_0 D_0 - \rho D = 0,$$

or, in a more extended form,

$$\rho_0(a, b, c, t_0) \frac{D(x_0, y_0, z_0)}{D(a, b, c)} = \rho(a, b, c, t) \frac{D(X, Y, Z)}{D(a, b, c)},$$

(1.10)

should take place. It is exactly this relation that is the desired continuity equation in Lagrangian variables. For an incompressible fluid ($\rho_t = 0$), the continuity equation takes the form

$$\frac{D(X, Y, Z)}{D(a, b, c)} = D_0(a, b, c).$$

(1.11)

If the Lagrangian variables coincide with the initial coordinates of the fluid particle (see Eq. (1.5)), then the right-hand side of Eq. (1.11) is equal to unity. In the general case, the Jacobian D does not depend on time, being only a function of Lagrangian coordinates.

The continuity equation can be written differently. Let us introduce the Jacobi matrix

$$\hat{R} = \begin{pmatrix} X_a & X_b & X_c \\ Y_a & Y_b & Y_c \\ Z_a & Z_b & Z_c \end{pmatrix}, \tag{1.12}$$

the elements of which are derivatives of the current coordinates of the particles with respect to Lagrangian variables. Like the functions X, Y, and Z themselves, these elements depend on Lagrangian coordinates and time. If the dependence of X, Y, and Z on a, b, and c is linear, then the Jacobi matrix consists only of functions of time.

The continuity equation in this approach is formulated as the condition for the determinant of the $\hat{R}$ matrix to be constant in time:

$$\det \hat{R} = \det \hat{R}_0, \quad \hat{R}_0 = \hat{R}|_{t=t_0}. \tag{1.13}$$

If relations (1.5) are fulfilled, then $\hat{R}_0$ is a unit matrix.

It should be emphasized that the continuity equation in Lagrangian variables has a very sophisticated form. Even if we consider the motion of a homogeneous incompressible fluid (1.11), it has cubic nonlinearity in three-dimensional geometry and quadratic nonlinearity in two-dimensional geometry. In Eulerian variables, the continuity Eq. (1.1) in both cases is linear and meets the requirement that the velocity field be solenoidal. For a stratified and compressible fluid, Eq. (1.10) becomes even more inaccessible for analysis.

As a historical note, we point out that the continuity equation was first derived by Lagrange for the case of an incompressible fluid and when condition (1.5) is fulfilled. In this case, $D_0 = 1$ in Eq. (1.11). Technically, it seemed to be as follows: Lagrange wrote down the equation $div\vec{V} = 0$ and then moved on to Lagrangian variables [5].

1.3. Momentum Equations

1.3.1. *Vector and coordinate forms*

To obtain Lagrange momentum equations, we write Newton's second law for an individual fluid particle:

$$\overrightarrow{R_{tt}} = \frac{\partial^2 \vec{R}}{\partial t^2} = -\frac{1}{\rho}\nabla p + \vec{f}, \tag{1.14}$$

where $\vec{R} = \{X, Y, Z\}$. On the right-hand sides of Eq. (1.14), differentiation occurs with respect to the unknown functions X, Y, and Z. To eliminate this, we multiply Eq. (1.14) in a scalar way by a vector tangent to one of the fluid coordinate curves, e.g., $\vec{R_a}$. Then, we obtain

$$\vec{R_{tt}}\vec{R_a} = -\frac{1}{\rho}\nabla p\vec{R_a} + (\vec{f} \cdot \vec{R_a}) = -\frac{1}{\rho}p_a + (\vec{f} \cdot \vec{R_a}). \qquad (1.15.1)$$

This is one of the three momentum equations. Similarly, we obtain two other equations

$$\vec{R_{tt}}\vec{R_b} = -\frac{1}{\rho}p_b + (\vec{f} \cdot \vec{R_b}); \qquad (1.15.2)$$

$$\vec{R_{tt}}\vec{R_c} = -\frac{1}{\rho}p_c + (\vec{f} \cdot \vec{R_c}). \qquad (1.15.3)$$

Equations (1.15.1–1.15.3) are the vector form of the momentum equations. In coordinate form, they look as follows:

$$X_{tt}X_a + Y_{tt}Y_a + Z_{tt}Z_a = -\frac{1}{\rho}p_a + (f_1 \cdot X_a + f_2 \cdot Y_a + f_3 \cdot Z_a),$$

$$X_{tt}X_b + Y_{tt}Y_b + Z_{tt}Z_b = -\frac{1}{\rho}p_b + (f_1 \cdot X_b + f_2 \cdot Y_b + f_3 \cdot Z_b),$$

$$X_{tt}X_c + Y_{tt}Y_c + Z_{tt}Z_c = -\frac{1}{\rho}p_c + (f_1 \cdot X_c + f_2 \cdot Y_c + f_3 \cdot Z_c).$$

$$(1.16)$$

Here, f_1, f_2, and f_3 are the components of the force $\vec{f}$ in projection on the Cartesian axes x, y, and z, respectively. Along with continuity Eq. (1.10), Eq. (1.16) forms a complete system of equations of an inviscid fluid in Lagrangian variables.

Using the Jacobi matrix, Eq. (1.16) is written as follows [7]:

$$\hat{R}^T(\vec{R}_{tt} - \vec{f}(\vec{R}, t)) = -\frac{1}{0}\nabla_{\vec{a}}p. \qquad (1.17)$$

The superscript "T" denotes the transpose operation, the coordinate components of the vector in the parentheses should be considered elements of the column vector, and the right-hand side uses the gradient notation for Lagrangian variables $\vec{a} = \{a, b, c\}$.

In Lagrangian variables, the momentum equations are quadratically nonlinear, just like the Euler equations. From this point of view, it cannot be said that either of the two approaches to describing fluid dynamics has advantages.

Equation (1.16) can also be derived due to Hamilton's variational principle. This is exactly what Lagrange did [5]. These equations were derived in modern notation [8] for an incompressible fluid and for a compressible fluid [9, 10] (see also [11]).

1.3.2. *Weber momentum equations*

Assume that the external force is potential, i.e.,

$$\vec{f} = -\nabla H.$$

Here, the gradient of the variables x, y, and z from the potential H is written on the right-hand side. For such a force, Eq. (1.16) can be represented as follows:

$$X_{tt}X_a + Y_{tt}Y_a + Z_{tt}Z_a = -\frac{1}{\rho}p_a - H_a,$$

$$X_{tt}X_b + Y_{tt}Y_b + Z_{tt}Z_b = -\frac{1}{\rho}p_b - H_b, \qquad (1.18)$$

$$X_{tt}X_c + Y_{tt}Y_c + Z_{tt}Z_c = -\frac{1}{\rho}p_c - H_c.$$

We integrate these equations over time in the range from t_0 to t. Taking into account that

$$\int_{t_0}^{t} X_{tt}X_a dt = X_t X_a|_{t_0}^{t} - \int_{t_0}^{t} X_t X_{ta} dt$$

$$= X_t X_a - X_{t0}\frac{\partial x_0}{\partial a} - \frac{1}{2}\frac{\partial}{\partial a}\int_{t_0}^{t} X_t^2 dt, \qquad (1.19)$$

where X_{t0} is the initial value of the velocity component of the fluid particle (a, b, c) in the x direction, and assuming

$$\Phi = \int_{t_0}^{t}\left[\int \frac{dp}{\rho} + H - \frac{1}{2}(X_t^2 + Y_t^2 + Z_t^2)\right]dt, \qquad (1.20)$$

we have

$$X_t X_a + Y_t Y_a + Z_t Z_a - (\overrightarrow{R_{t0}} \cdot \overrightarrow{R_{0a}}) = -\mathbf{\Phi}_a,$$

$$X_t X_b + Y_t Y_b + Z_t Z_b - (\overrightarrow{R_{t0}} \cdot \overrightarrow{R_{0b}}) = -\mathbf{\Phi}_b, \text{ and} \qquad (1.21)$$

$$X_t X_c + Y_t Y_c + Z_t Z_{ca} - (\overrightarrow{R_{t0}} \cdot \overrightarrow{R_{0c}}) = -\mathbf{\Phi}_{ca}.$$

Here, $\vec{R}_{t0}$ is the vector of the initial velocity of the fluid particle and $\vec{R}_0$ is its initial radius vector. When deriving Eq. (1.20), it was assumed that the fluid density is either constant or is a function of pressure (i.e., the fluid is barotropic).

These three equations, together with the equation

$$\frac{\partial \Phi}{\partial t} = \int \frac{dp}{\rho} + H - \frac{1}{2}(X_t^2 + Y_t^2 + Z_t^2) \qquad (1.22)$$

and the continuity equation, form a system of partial differential equations to determine the five unknown quantities, X, Y, Z, p, and Φ. The system of equations was first proposed by Heinrich Martin Weber [2, 12].

1.3.3. *Cauchy momentum equations*

In the case of a potential external force, one can obtain another well-known form of the momentum equations. For this purpose, we take the cross derivatives for each pair of Eqs. (1.18) and subtract one from the other. The right-hand sides with pressure derivatives (hereafter, we assume the fluid density to be constant everywhere) and potential force will disappear, and the left-hand sides will appear as time derivatives. Upon integration over time, we obtain three equations [2]:

$$X_{tb} X_c - X_{tc} X_b + Y_{tb} Y_c - Y_{tc} Y_b + Z_{tb} Z_c - Z_{tc} Z_b = S_1(a, b, c);$$

$$X_{tc} X_a - X_{ta} X_c + Y_{tc} Y_a - Y_{ta} Y_c + Z_{tc} Z_a - Z_{ta} Z_c = S_2(a, b, c);$$

$$X_{ta} X_b - X_{tb} X_a + Y_{ta} Y_b - Y_{tb} Y_a + Z_{ta} Z_b - Z_{tb} Z_a = S_3(a, b, c).$$
$$(1.23)$$

Here, S_1, S_2, and S_3 are arbitrary functions of Lagrangian variables (momentum integrals). Note that as a result, it was possible to eliminate pressure without increasing the order of the equations. Equation (1.23) was first formulated by Cauchy [13, 14]. For this work,

Cauchy received an award from the Paris Academy of Sciences in 1815. Stokes called the functions S_1, S_2, and S_3 Cauchy integrals [14, 15]. In a study [16], the term Cauchy invariants was proposed for them. This terminology has successfully taken root [4, 14, 16–19]. Equation (1.23) will remain valid for a barotropic fluid.

By directly differentiating Eq. (1.23), one can verify that the derivatives of the Cauchy invariants S_1, S_2, and S_3 with respect to the Lagrangian variables are connected by the following relation:

$$\frac{\partial S_1}{\partial a} + \frac{\partial S_2}{\partial b} + \frac{\partial S_3}{\partial c} = 0.$$

This means that they cannot be specified arbitrarily. We introduce a vector of Cauchy invariants $\vec{S}\{S_1, S_2, S_3\}$, so that

$$\vec{S} = S_1 \vec{a_0} + S_1 \vec{b_0} + S_1 \vec{c_0},$$

where $\vec{a_0}$, $\vec{b_0}$, and $\vec{c_0}$ are unit vectors along the corresponding axes. The divergence of this vector should vanish:

$$div_{\vec{a}}\vec{S} = 0.$$

System (1.23) can be rewritten in a more compact form:

$$\vec{S} = rot_{\vec{a}} \sum_{i=1}^{3} (\vec{R_t}\vec{R_{a_l}})\vec{a_{l0}}; \tag{1.24}$$

here, we used the notation $\vec{a} = \{a_1, a_2, a_3\} = \{a, b, c\}$, and the index of the rotor operation means it is performed in Lagrangian variables.

Equation (1.24) can also be obtained from Weber equations. We find the cross derivatives with respect to Lagrangian variables for each pair of Eqs. (1.21) and subtract one from the other. In this case, the connection between the Cauchy invariants and the initial conditions specified by the vectors $\vec{R_0}$ and $\vec{R_{t0}}$ is determined:

$$S_1 = \frac{\partial}{\partial b}(\vec{R_{t0}}\vec{R_{0c}}) - \frac{\partial}{\partial c}(\vec{R_{t0}}\vec{R_{0b}}),$$

$$S_2 = \frac{\partial}{\partial c}(\vec{R_{t0}}\vec{R_{0a}}) - \frac{\partial}{\partial a}(\vec{R_{t0}}\vec{R_{0c}}), \tag{1.25}$$

$$S_3 = \frac{\partial}{\partial a}(\vec{R_{t0}}\vec{R_{0b}}) - \frac{\partial}{\partial c}(\vec{R_{t0}}\vec{R_{0a}}).$$

1.3.4. *Matrix form of momentum equations*

By direct checking, it can be verified that by using the Jacobi matrix (1.12), the system of Eq. (1.24) can be reduced to a single matrix equation [16, 20]:

$$\hat{R}_t^T \hat{R} - \hat{R}^T \hat{R}_t = \hat{S}; \quad \hat{S} = \begin{pmatrix} 0 & S_3 & -S_2 \\ -S_3 & 0 & S_1 \\ S_2 & -S_1 & 0 \end{pmatrix}. \tag{1.26}$$

The right-hand side of this equation is a time-independent antisymmetric matrix $\hat{S}$, the non-zero elements of which are equal to the Cauchy invariants with a plus or minus sign. Together with the continuity Eq. (1.13), it forms a system of matrix equations for the fluid dynamics of an ideal incompressible fluid. The condition that the matrix $\hat{R}$ is not arbitrary should be added to these equations. From its definition (1.12), it follows that the cross derivatives of the column elements must be equal to each other:

$$\frac{\partial R_{ik}}{\partial a_n} = \frac{\partial R_{in}}{\partial a_k}. \tag{1.27}$$

Here, all indices range from values of 1 to 3.

Let the change in the infinitesimal fluid element $\overrightarrow{dR}$ $\{dX, dY, dZ\}$ correspond to the increment of Lagrangian coordinates $\overrightarrow{da}$ $\{da, db, dc\}$. Assuming that $\overrightarrow{dR}$ and $\overrightarrow{da}$ are vector columns, the following formula is valid:

$$\overrightarrow{dR} = \hat{R}\overrightarrow{da},$$

i.e., the $\hat{R}$ matrix relates infinitesimal changes in physical (Eulerian) and Lagrangian spaces. We emphasize that the systems (1.13), (1.26), and (1.27) should not be considered as a new closed approach to describing the motion of an incompressible fluid based on the Jacobi matrix. It is just an elegant and convenient way to write fluid dynamics equations. Even formally, such a statement would be incorrect since momentum Eq. (1.26), along with the $\hat{R}$ matrix itself, also contains a matrix transposed to it, and the continuity equation includes a determinant calculation operation that belongs to matrix elements, i.e., derivatives of trajectory coordinates with respect to Lagrangian variables.

1.4. Vorticity and Cauchy Invariants

Vorticity $\vec{\Omega}(\vec{r}, t)$ is defined as the curl of velocity

$$\vec{\Omega}(\vec{r}, t) = \{\Omega_x, \Omega_y, \Omega_z\} = rot\vec{V}$$
$$= \left\{\frac{\partial V_z}{\partial y} - \frac{\partial V_y}{\partial z}, \frac{\partial V_x}{\partial z} - \frac{\partial V_z}{\partial x}, \frac{\partial V_y}{\partial x} - \frac{\partial V_x}{\partial y}\right\}, \quad (1.28)$$

where $\vec{r}$ $\{x, y, z\}$ is the radius vector of the observation point and $\vec{V}$ $\{V_x, V_y, V_z\}$ is the velocity vector with components along the x, y, and z axes. Subscripts of the values in Eq. (1.28) denote projections on the corresponding axis. A line at any point at which the velocity vector is directed tangentially is called a vortex line. If one takes a closed line and draws a vortex line through each of its points, then the totality of all such vortex lines forms a vortex tube. Vorticity, along with the velocity, enters into the fluid momentum equation, which in the inviscid limit can also be written as

$$\frac{\partial \vec{V}}{\partial t} + \nabla \frac{1}{2} V^2 - \vec{V} \times \vec{\Omega} = -\frac{1}{\rho} \nabla p + \vec{f}. \quad (1.29)$$

Its difference from the classical Euler Eq. (1.2) is that the nonlinear term $(\vec{V} \cdot \nabla)\vec{V}$ is represented in a different form using vector analysis formulas. If we assume that the fluid is barotropic and apply the rotor-taking operation to Eq. (1.29), we will ultimately obtain the following equation:

$$\frac{d\vec{\Omega}}{dt} - (\vec{\Omega} \cdot \nabla)\vec{V} + \vec{\Omega} div \vec{V} = 0. \quad (1.30)$$

For an incompressible fluid, the last term on the left-hand side is zero. Equation (1.30) no longer contains pressure, and the rate of change of vorticity depends only on the instantaneous local velocity fields and the vorticity itself.

In his 1858 treatise [21], Helmholtz adduced proof for two theorems [22].

Theorem 1. *Let* (i) *the fluid be ideal,* (ii) *the force $\vec{f}$ have a potential, and* (iii) *the fluid density be a function of pressure. Then the fluid particles forming a vortex line at some moment will always form a vortex line.*

Theorem 2. *Under the assumptions of Theorem 1, the intensity of any vortex tube remains the same throughout the entire process of motion.*

Equations (1.28)–(1.30) represent the concept of vorticity in terms of the Euler description. In Lagrangian coordinates, the formulas for the components of the vortex vector are written as follows:

$$
\begin{aligned}
\Omega_x &= \frac{\partial Z_t}{\partial Y} - \frac{\partial Y_t}{\partial Z} = \frac{D(X, Z_t, Z)}{D(X, Y, Z)} - \frac{D(X, Y_t, Y)}{D(X, Z, Y)} \\
&= \frac{1}{D_0}\left[\frac{D(Z_t, Z, X)}{D(a, b, c)} + \frac{D(Y_t, Y, X)}{D(a, b, c)}\right]; \\
\Omega_y &= \frac{\partial X_t}{\partial Z} - \frac{\partial Z_t}{\partial X} = \frac{D(X, Y, X_t)}{D(X, Y, Z)} - \frac{D(Z_t, Y, Z)}{D(X, Y, Z)} \\
&= \frac{1}{D_0}\left[\frac{D(Z_t, Z, Y)}{D(a, b, c)} + \frac{D(X_t, X, Y)}{D(a, b, c)}\right]; \\
\Omega_z &= \frac{\partial Y_t}{\partial Z} - \frac{\partial X_t}{\partial X} = \frac{D(Y_t, Y, Z)}{D(X, Y, Z)} - \frac{D(X, X_t, Z)}{D(X, Y, Z)} \\
&= \frac{1}{D_0}\left[\frac{D(Y_t, Y, Z)}{D(a, b, c)} + \frac{D(X_t, X, Z)}{D(a, b, c)}\right],
\end{aligned}
\tag{1.31}
$$

where the sequence of elementary transformations related to Jacobian "algebra" is deliberately indicated.

For two-dimensional flows, X and Y depend only on a, b, and t and $Z = c$. As can be seen from Eq. (1.24), the Cauchy invariants S_1 and S_2 in this case are equal to zero. Similarly, the vortex components Ω_x and Ω_y are also equal to zero (see Eq. (1.31)). The third vortex component Ω_z has the form

$$
\Omega_z = \frac{1}{D_0}\left[\frac{D(X_t, X)}{D(a, b)} + \frac{D(Y_t, Y)}{D(a, b)}\right] = \frac{S_3(a, b)}{D_0(a, b)}.
\tag{1.32}
$$

The component Ω_z is a function of Lagrangian coordinates and is independent of time, which means that during two-dimensional motion, the vorticity of fluid particles is preserved. The invariant S_3 is generally proportional to Ω_z and is equal to it if $D_0 = 1$. We now address three-dimensional motion.

Ideal Fluid Equations

Transform Eq. (1.24). Let us present the derivatives included in the equation (from the functions X_t, Y_t, and Z_t with respect to the coordinates a, b, and c) via the derivatives from the same functions with respect to the variables X, Y, and Z. For example,

$$X_{ta} = \frac{\partial X_t}{\partial X} X_a + \frac{\partial X_t}{\partial Y} Y_a + \frac{\partial X_t}{\partial Z} Z_a, \text{ etc.} \tag{1.33}$$

Substituting equations of the form (1.33) into Eq. (1.24), we obtain the following system:

$$\Omega_x \frac{D(Y,Z)}{D(b,c)} + \Omega_y \frac{D(Z,X)}{D(b,c)} + \Omega_z \frac{D(X,Y)}{D(b,c)} = S_1;$$

$$\Omega_x \frac{D(Y,Z)}{D(c,a)} + \Omega_y \frac{D(Z,X)}{D(c,a)} + \Omega_z \frac{D(X,Y)}{D(c,a)} = S_2; \tag{1.34}$$

$$\Omega_x \frac{D(Y,Z)}{D(a,b)} + \Omega_y \frac{D(Z,X)}{D(a,b)} + \Omega_z \frac{D(X,Y)}{D(a,b)} = S_3.$$

These equations establish a connection between the components of the vorticity vector and the Cauchy invariants. However, a simpler representation is possible for system (1.34). Let us multiply the first of its equations by X_a, the second by X_b, the third by X_c, and sum them up. Taking into account continuity Eq. (1.11), we obtain the following expression:

$$\Omega_x = D_0^{-1}(S_1 X_a + S_2 X_b + S_3 X_c). \tag{1.35}$$

In a similar way (by multiplying Eq. (1.34) by the corresponding derivatives of the functions Y and Z and summing), we find expressions for the other two vorticity components:

$$\Omega_\nu = D_0^{-1}(S_1 Y_a + S_2 Y_p + S_3 Y_c) \tag{1.36}$$

and

$$\Omega_z = D_0^{-1}(S_1 Z_a + S_2 Z_b + S_3 Z_c). \tag{1.37}$$

Equations 1.35)–(1.37) can be rewritten in vector form,

$$\vec{\Omega} = D_0^{-1}(S_1 \vec{R}_a + S_2 \vec{R}_b + S_3 \vec{R}_c), \tag{1.38}$$

18 *Analytical Fluid Dynamics in Lagrangian Variables*

or, using the Jacobi matrix, in the form

$$\vec{\Omega} = \frac{\hat{R}}{\det \hat{R}_0} \vec{S}. \tag{1.39}$$

Here again, $\vec{\Omega}(\Omega_x, \Omega_y, \Omega_z)$ and $\vec{S}$ are meant to be column vectors. In particular, when $t = 0$, we have

$$\vec{\Omega}_0 = \frac{\hat{R}_0}{D_0} \vec{S}. \tag{1.40}$$

If the initial coordinates X_0, Y_0, and Z_0 of fluid particles are chosen as Lagrangian variables, i.e., if Eq. (1.5) for $t_0 = 0$ is valid, then the $\hat{R}_0$ matrix coincides with the identity matrix and the relations

$$S_1 = \Omega_{x0}, \quad S_2 = \Omega_{y0}, \quad \text{and} \quad S_3 = \Omega_{z0} \tag{1.41}$$

are fulfilled.

This result was obtained by Cauchy [2,13]. It can also be obtained from Eq. (1.25). Both Helmholtz theorems obviously follow from the fundamental fact of the connection between Cauchy invariants and the initial vorticity.

In view of Eqs. (1.41) and (1.39) will be rewritten as

$$\vec{\Omega} = \hat{R}\vec{\Omega}_0, \tag{1.42}$$

or, in coordinate form, as

$$\Omega_i = \Omega_{0j} \frac{\partial R_i}{\partial a_j}. \tag{1.43}$$

This relation is called the Cauchy vorticity formula. The latter is obviously equivalent to Eqs. (1.35)–(1.37) for the case $D_0 = 1$ and $\vec{S} = \vec{\Omega}_0$.

1.5. Physical Meaning of Cauchy Invariants

Let us show, based on a study [2], that Eq. (1.34) reflects the fundamental fact of the conservation of circulation in a closed loop. Let us choose a closed curve, which at the initial moment bounded, for example, the rectangle $\delta b \, \delta c$ (in the Lagrangian plane $a = $ const), and denote by A, B, and C the areas of its projections on the coordinate planes at time t. They will be equal to

$$A = |d\vec{Y} \times d\vec{Z}| = \frac{D(Y,Z)}{D(b,c)}\delta b\, \delta c,$$

$$B = |d\vec{Z} \times d\vec{X}|\frac{D(Z,X)}{D(b,c)}\delta b\, \delta c, \quad \text{and} \qquad (1.44)$$

$$C = |d\vec{X} \times d\vec{Y}| = \frac{D(X,Y)}{D(b,c)}\delta b\, \delta c,$$

respectively. Then, the first of the Eq. (1.34) will be written as follows:

$$\Omega_x A + \Omega_y B + \Omega_z C = S_1 \delta b\, \delta c. \qquad (1.45)$$

Thus, the flux of the vorticity vector through the area consisting of fluid particles in the $\delta b\, \delta c$ rectangle is independent of time. Accordingly, the circulation of speed around the loop bounding this area remains constant. The S_1 value has the meaning of the vorticity flux through a unit area in the plane of Lagrangian variables b and c. In other words, the independence of the S_1 value from time is related to the condition for maintaining speed circulation along an infinitesimal loop covering the $\delta b\, \delta c$ area. These conclusions can be generalized to any region consisting of a finite set of such areas in the $a = \text{const}$ plane and to any closed loop in it. The physical meaning of the other two invariants can be explained in a similar way.

Our reasoning is reversible, i.e., one can deduce the condition for maintaining speed circulation around a closed loop from the independence of S_1, S_2, and S_3 from time. Indeed, let us choose Lagrangian coordinates so that the particles forming a given fluid loop lie, for example, in the plane of the variables b and c. Then, due to the invariance of S_1, the circulation around this loop will remain constant.

The same results can be achieved more formally. Let us choose some surface Σ_0, bounded by the closed loop C_0, in the space of Lagrangian variables and apply Stokes' theorem to Eq. (1.24):

$$\iint_{\Sigma_0} \vec{S}d\vec{\sigma} = \int_{c_0} \sum_{i=1}^{3}\left(\vec{R_t}\vec{R_{a_l}}\right)\vec{a_{t0}} \cdot \vec{da} = \int_{c(t)} \vec{R_t} \cdot \vec{dR}. \qquad (1.46)$$

Here, $\vec{d\sigma}$ is an area $d\sigma$ element multiplied by the unit normal. The rightmost integral represents the circulation in the physical space

x, y, z around the loop $C(t)$ formed at the moment $t > t_0$ by the particles of the loop C_0. Since the leftmost integral is independent of time, the speed circulation will remain.

This result (within the framework of the Lagrangian approach) was first obtained by Hankel in 1861 [14, 23], eight years before Thomson (Lord Kelvin) came to it using the Eulerian description [24]. Based on this, Trusdell proposed to call the theorem of conservation of circulation the Hankel–Kelvin theorem [3, 14], but this idea was not supported.

Equation (1.44), demonstrating the fact of constant circulation along a fluid loop, can be written in a more elegant form. Let us consider Lagrangian variables a, b, and c as functions of Eulerian variables:

$$a = a(x, y, z), \quad b = b(x, y, z), \quad c = c(x, y, z),$$

Then, on solving the system of equations

$$a_x x_a + a_y y_a + a_z z_a = 1;$$

$$a_x x_b + a_y y_b + a_z z_b = 0;$$

$$a_x x_c + a_y y_c + a_z z_c = 0,$$

we find that

$$a_x = \frac{1}{D_0} \frac{D(y, z)}{D(b, c)}; \quad a_y = \frac{1}{D_0} \frac{D(z, x)}{D(b, c)}; \quad a_z = \frac{1}{D_0} \frac{D(x, y)}{D(b, c)}. \quad (1.47)$$

Comparing Eqs. (1.44), (1.45), and (1.47), we conclude that the equation

$$S_1 = D_0(\vec{\Omega} \nabla a) \quad (1.48)$$

is valid. Representations for the other two Cauchy invariants (S_2 and S_3) can be obtained in a similar way:

$$S_2 = D_0(\vec{\Omega} \nabla b), \quad S_3 = D_0(\vec{\Omega} \nabla c). \quad (1.49)$$

The structure of Eqs. (1.48) and (1.49) resembles Ertel's result that, when the flow is non-isentropic, for each moving particle, the related quantity $(\vec{\Omega} \nabla \varepsilon)/\rho$ where ε is entropy, remains constant [25, pp. 30–31].

Ideal Fluid Equations 21

At the end of this chapter, we show a result disclosing the deep nature of the existence of Cauchy invariants. Noether's theorem states that every continuous symmetry of a physical system corresponds to some conservation law. Homogeneity in time corresponds to the law of conservation of energy, homogeneity of space to the law of conservation of momentum, and isotropy of space to the law of conservation of angular momentum. Some papers [10, 26, 27] demonstrate that the invariance of the functions S_1, S_2, and S_3 is related to "relabeling symmetry." A detailed review of relevant studies, including additional references, can be found elsewhere [11]. The history of the discovery of this type of symmetry, and its connection with Cauchy invariants, is also described elsewhere [14].

The essence of relabeling symmetry is that arbitrary labeling of fluid particles is allowed, provided that the mass of the fluid particles (or volume, if the fluid is incompressible) is conserved. The description of fluid motion does not depend on the numbering of particles. From the point of view of common sense, this is quite natural since obtaining new solutions (and, accordingly, flows) just by using a different designation for particles would be very strange. In practice, this means that the initial positions of the fluid particles can always be chosen as Lagrangian variables. The Jacobian D_0 in the continuity equation will be equal to unity, which simplifies the study. This circumstance should be taken into account when justifying general results and during numerical simulation. Another point is that the Lagrangian form is not always convenient for the analytical study of equations. As we will see in the following, exact solutions are frequently so designed that the initial positions of the particles do not coincide with the chosen Lagrangian coordinates, and an attempt to analytically record a flow that satisfies condition (1.5), i.e., to do this using different labeling, is simply impossible to implement (due to the impossibility, for example, to solve transcendental equations).

1.6. Fluid Dynamics Equations in Curvilinear Coordinate Systems

All previous reasonings were carried out in Cartesian coordinates. To expand the scope of the description, we define the form of the fluid dynamics equations in cylindrical and spherical coordinates.

1.6.1. Cylindrical geometry

We consider the motion of fluid particles in the cylindrical coordinate system R, Φ, Z (here, R is the distance to the Z-axis) and choose the Lagrangian cylindrical coordinates r, φ, and z as markers of an individual particle so that the trajectories of the particles are determined by the system of equations

$$R = R(r, \varphi, z, t), \quad \Phi = \Phi(r, \varphi, z, t), \quad Z = Z(r, \varphi, z, t). \quad (1.50)$$

The continuity equation (1.10) in this case can be written in the form

$$\frac{\partial}{\partial t} \left[\rho(r, \varphi, z, t) \frac{D(X, Y, Z)}{D(R, \Phi, Z)} \frac{D(R, \Phi, Z)}{D(r, \varphi, z)} \frac{D(r, \varphi, z)}{D(a, b, c)} \right]$$
$$= \frac{1}{r} \frac{\partial}{\partial t} \left[\rho R \frac{D(R, \Phi, Z)}{D(r, \varphi, z)} \right] = 0. \quad (1.51)$$

Discarding the factor r^{-1}, we write Eq. (1.51) in its final form:

$$\frac{\partial}{\partial t}(\rho R D) = 0, \quad D = \frac{D(R, \Phi, Z)}{D(r, \varphi, z)}. \quad (1.52)$$

The momentum equations are written as follows:

$$R_{tt} - R\Phi_t^2 = -\frac{1}{\rho} \frac{\partial p}{\partial R}; \quad (1.53)$$

$$R\Phi_{tt} + 2R_t\Phi_t = -\frac{1}{\rho R} \frac{\partial p}{\partial \Phi}; \quad (1.54)$$

$$Z_{tt} = -\frac{1}{\rho} \frac{\partial p}{\partial Z}. \quad (1.55)$$

The derivation of expressions for the acceleration components is given in Appendix 1. Systems (1.53)–(1.55) have a mixed (Lagrangian–Eulerian) form since on the right-hand sides the differentiation is performed with respect to Eulerian variables. To write this system in Lagrangian representation, the following substitutions should be made:

$$\frac{\partial p}{\partial R} = D^{-1} \frac{D(p, \Phi, Z)}{D(r, \varphi, z)}; \quad \frac{\partial p}{\partial \Phi} = D^{-1} \frac{D(p, Z, R)}{D(r, \varphi, z)};$$
$$\frac{\partial p}{\partial Z} = D^{-1} \frac{D(p, R, \Phi)}{D(r, \varphi, z)}. \quad (1.56)$$

Ideal Fluid Equations

In general, both the continuity equation (1.52) and all momentum equations (1.53)–(1.56) have an awesome form even in the case of a homogeneous incompressible fluid.

1.6.2. *Spherical geometry*

We study the fluid motion in a spherical coordinate system R, Θ, Φ, where R is the distance to the origin of coordinates, and Θ and Φ are the polar and azimuthal angles, respectively. The Lagrangian coordinates r, θ, and φ have a similar meaning. The trajectories of fluid particles are determined by the relations

$$R = R(r, \Theta, \varphi, t), \quad \Theta = \Theta(r, \Theta, \varphi, t), \quad \text{and} \quad \Phi = \Phi(r, \Theta, \varphi, t).$$

$$(1.57)$$

According to the same scheme as in Section 1.6.1, we find a system of fluid dynamics equations in spherical coordinates in Lagrangian form:

$$\frac{\partial}{\partial t}(\rho R^2 D \sin \theta) = 0; \quad D = \frac{D(R, \Theta, \Phi)}{D(r, \theta, \varphi)}, \tag{1.58}$$

$$R_{tt} - R\Theta_t^2 - R\Phi_t^2 \sin^2 \theta = -\frac{1}{\rho D} \frac{D(p, \Theta, \Phi)}{D(r, \theta, \varphi)}, \tag{1.59}$$

$$R\Theta_{tt} + 2R_t\Theta_t - R\Phi_t^2 \sin \Theta \cos \Theta = -\frac{1}{\rho R D} \frac{D(p, \Phi, R)}{D(r, \theta, \varphi)}, \tag{1.60}$$

$$R\Phi_{tt} \sin \Theta + 2R_t\Phi_t \sin \Theta + 2R\Phi_t\Theta_t \cos \Theta$$
$$= -\frac{1}{\rho R D \sin \Theta} \frac{D(p, R, \Theta)}{D(r, \theta, \varphi)}. \tag{1.61}$$

Calculations for acceleration components are given in Appendix 2. Systems (1.58)–(1.61), as well as systems (1.52)–(1.56), are incomparably more complex than the corresponding equations in Eulerian variables.

References

[1] Euler, L. (1757). Principes généraux du movement des fluids. Mémories de l'Académie royale des sciencies et belles letters. *Berlin*,

11(1755), pp. 274–315; Frish, U. (2008) General Principles of the Motion of Fluids, *arXiv:0802.2383v1*.

[2] Lamb, H. (1932). *Hydrodynamics*, 6th edn. (Cambridge University Press, Cambridge).

[3] Truesdell, C. (1954). *The Kinematics of Vorticity* (Indiana University Press, Bloomington).

[4] Bennett, A. (2006). *Lagrangian Fluid Mechanics* (Cambridge University Press, New York).

[5] Lagrange, J.L. (1997). *Analytical Mechanics* (Springer, Dordrecht).

[6] Monin, A.S. and Yaglom, A. M. (1971). *Statistical Fluid Mechanics: Mechanics of Turbulence*, Vol. 1 (MIT Press, Cambridge).

[7] Ovsyannikov, L.V. (1967). *General Equations and Examples. The Problem of Unsteady Fluid Motion with a Free Boundary* [in Russian], (Nauka, Novosobirsk).

[8] Herivel, J.W. (1955). The derivation of the equations of motion of an ideal fluid by Hamilton's principle. *Proc. Cambridge Philos. Soc.*, 51, pp. 344–349.

[9] Serrin, J. (1959). *Mathematical Principles of Classical Fluid Mechanics. Handbuch der Physic VIII-I*, pp. 125–263 (Springer-Verlag, Berlin).

[10] Eckart, C. (1960). Variational principles of hydrodynamics. *Phys. Fluids*, 3, pp. 421–427.

[11] Salmon, R. (1988). Hamilton fluid mechanics. *Ann. Rev. Fluid Mech.*, 20(1), pp. 225–256.

[12] Weber, H.M. (1898). Über eine Transformation der hydrodynamischen Gleichunge. *J. für die Reine und Ang. Math. (Crelle)*, 68, pp. 286–292.

[13] Cauchy, Augustin-Louis. (1815/1827). 'Théorie de la propagation des ondes à la surface d'un fluide pesant d'une profondeur indéfinie — Prix d'analyse mathématique remporté par M. Augustin-Louis Cauchy, ingénieur des Ponts et Chaussées. (Concours de 1815).' Mémoires présentés par divers savans à l'Académie royale des sciences de l'Institut de France et imprimés par son ordre. Sciences mathématiques et physiques. Tome I, imprimé par autorisation du Roi à l'Imprimerie royale, pp. 5–318.

[14] Frisch, U. and Villone, B. (2014). Cauchy's almost forgotten Lagrangian formulation of the Euler equation for 3D incompressible flow. *Eur. Phys. J. H*, 39, pp. 325–351.

[15] Stokes, G.G. (1848). Notes on hydrodynamics. IV demonstration of a fundamental theorem. *The Cambridge and Dublin Mathematical Journal*, 3, pp. 209–219.

[16] Abrashkin, A.A., Zen'kovich, D.A. and Yakubovich, E.I. (1996). Matrix formulation of hydrodynamics and extension of Ptolemaic flows to three-dimensional motions. *Radiophys. And Quant. Electronics.*, 39(6), pp. 518–526.

[17] Zakharov, V.E. and Kuznetsov, E.A. (1997). Hamiltonian formalism for nonlinear waves. *Phys.-Usp.*, 40(11), pp. 1087–1116.

[18] Kuznetsov, E.A. (2006). Vortex line representation for the hydrodynamic type equations. *J. Nonl. Math. Phys.*, 13(1), pp. 64–80.

[19] Besse N. and Frisch, U. (2017). Geometric formulation of the Cauchy invariants for incompressible Euler flow in flat and curved spaces. *J. Fluid Mech.*, 825, pp. 412–478.

[20] Abrashkin, A.A., Zen'kovich, D.A. and Yakubovich, E.I. (1997). The Study of Three-dimensional Vortex Flows Using Matrix Equations of Hydrodynamics. *Phys.- Dokl.*, 42(12), pp. 687–690.

[21] Helmholtz, H. (1858). Über Integrale der hydrodynamischen Gleichungen, welche den Wirbelbewegungen entsprechen. *Journal für die reine und angewandte Mathematik*, 55, pp. 25–55. Translated into English by Tait, P.G. (1867). On integrals of the hydrodynamical equations, which express vortex motion. *The London, Edinburgh, and Dublin Philosophical Magazine*, suppl. to Vol. XXXIII, pp. 485–512.

[22] Kochin, N.E., Kibel', I.A. and Roze, N.V. (1964). *Theoretical Hydromechanics*, Vol. 1 (Interscience Publ., New York).

[23] Hankel, H. (1861). Zur allgemeinen Theorie der Bewegung der Flüssigkeiten. Eine von der philosophischen Facultät der Georgia Augusta am 4. Juni 1861 gekrönte Preisschrift, Göttingen. Printed by Dieterichschen Univ.-Buchdruckerei (W. Fr. Kaestner).

[24] Thomson, W. (Lord Kelvin). (1869). On vortex motion. *Transactions of the Royal Society of Edinburgh*, 25, pp. 217–260.

[25] Landau, L.D. and Lifshitz, E.M. (1959). *Fluid Mechanics* (Pergamon Press. London).

[26] Newcomb, W.A. (1967). Exchange invariance in fluid systems. *Proc. Symp. Appl. Math.*, 18, pp. 152–161.

[27] Eckart, C. (1963). Some transformation of the hydrodynamic equations. *Phys. Fluids*, 6(8), pp. 1037–1041.

Chapter 2

Potential Motion

> "The main assumption that played a historical
> role in bringing fluid dynamics closer to specific
> applications was the assumption of the absence
> of vorticity in a moving fluid."
>
> — L.G. Loitsyanskiy [1]

2.1. Flow Potentiality Conditions

If at some initial time there is no vorticity in the fluid, then the
quantities S_1, S_2, and S_3 are equal to zero. Due to the fact that
Cauchy invariants are momentum integrals, they will remain zero
at the next moments of time. It follows from Eqs. (1.48) and (1.49)
that the flow vorticity will be zero as well. Thus, Lagrange's theorem
is proved, i.e., vorticity will not arise in an ideal homogeneous (or
barotropic) fluid under the action of potential forces unless vortices
take place in the fluid at the initial time.

The necessary and sufficient condition for vortex-free (potential)
motion can be formulated using two equalities:

$$rot_{\vec{a}}\vec{V}_0 = 0 \quad \text{and} \quad rot_{\vec{r}}\vec{f} = 0.$$

The first of them is the condition of absence of the vorticity at the
initial time (it is taken into account that $\vec{R}_0 = \vec{a}$) and the second
equality is the external force potentiality condition (the subscript
$\vec{r}$ means that the operation is performed in the coordinate space
x, y, z).

The momentum equation for the potential flow has the form

$$\hat{R}_t^T \hat{R} - \hat{R}^T \hat{R}_t = 0. \tag{2.1}$$

In the coordinate representation, this relation includes three equations of form (1.23) with only zero right-hand sides. If we take into account that, in addition to them, it is also necessary to satisfy continuity equation (1.10) to describe the flow, it becomes clear that the analytical theory of potential flows is a very non-trivial problem.

With the Eulerian approach, the study is significantly simplified by introducing the flow potential. If $\varphi(\vec{r}, t)$ is the flow potential, then

$$\vec{V}(x, y, z, t) = \vec{r}_t = \nabla_{\vec{r}} \varphi. \tag{2.2}$$

Assume that $D_0 = \det \hat{R} = 1$. Then, with allowance for the fact that the gradients from an arbitrary continuous function ϕ over Lagrangian and Eulerian coordinates will be connected by the relation

$$\nabla_{\vec{a}} \varphi = \hat{R}^T \nabla_{\vec{r}} \varphi,$$

we obtain from Eq. (2.2) the following equality [2]:

$$\hat{R}^T \vec{r}_t = \nabla_{\vec{a}} \varphi.$$

It follows from this relation that in Lagrangian variables, the velocity vector of the potential flow can no longer always be represented by the scalar field gradient ($\vec{r}_t = \hat{R}^{T^{-1}} \nabla_{\vec{a}} \varphi$), and it is time to state that an accurate Lagrangian description can be given only in an extremely limited number of cases. This is a major drawback of this method, and it is exactly in this area that this method is seriously inferior to the Eulerian approach.

The following are some examples where it is still possible to obtain explicit analytical expressions.

2.2. Flow Near the Critical Point

The point of the fluid dynamic flow, where the velocity turns to zero, is called critical. It can occur inside the fluid or at its boundary. Let us assume for simplicity that the fluid is unbounded and the critical

Potential Motion 29

point O itself coincides with the origin of coordinates x, y, and z. In Eulerian variables up to a small order of r^2, the fluid motion near the critical point is described by the velocity field [3]

$$V_x = \alpha x, \quad V_y = \beta y, \quad V_z = -(\alpha + \beta)z, \qquad (2.3)$$

where α and β are constants related to the flow in the vicinity of point O. This velocity field is solenoidal and potential. The coordinates of the fluid particle trajectories for this flow are obtained from Eq. (2.3) by simple integration and can be written as follows:

$$X = ae^{\alpha t}, \quad Y = be^{\beta t}, \quad Z = ce^{-(\alpha + \beta)t}. \qquad (2.4)$$

Here, a, b, and c are the initial coordinates of the fluid particle at the time $t = 0$. Equation (2.4) is a Lagrangian representation of the spatial potential flow near the critical point. It is easy to make sure that it satisfies continuity equation (1.11), so that $D_0 = 1$. It follows from Eq. (1.23) that all the Cauchy invariants are zero, which means that the flow is potential.

2.2.1.   *Two-dimensional motion*

If the flow occurs in planes parallel to the XOY plane, one should set $Z = c$, i.e., assume $\alpha = -\beta$. In this case, the motion is determined by the formulas

$$X = ae^{\alpha t} \quad \text{and} \quad Y = be^{-\alpha t}. \qquad (2.5)$$

It is easy to verify that these formulas satisfy two-dimensional equations for potential motion,

$$X_a Y_b - X_b Y_a = 1, \qquad (2.6)$$

$$X_{ta} X_b - X_{tb} X_a + Y_{ta} Y_b - Y_{tb} Y_a = 0. \qquad (2.7)$$

It can be seen from Eq. (2.5) that the equality $XY = \text{const}$ is fulfilled for the coordinates of the trajectories. This means that fluid particles move along hyperbolas. In the Eulerian representation, the flow (2.5) corresponds to the stream function $\psi = \alpha xy$. The trajectories of fluid particles coincide with the streamlines $\psi = \text{const}$, as it should be for a steady flow. For those who are accustomed to the Eulerian description, the Lagrangian representation looks unusual at first glance. Equation (2.5) contains time, and it is difficult to immediately say that it describes a steady flow.

Consider one more example. Expressions

$$X = a\,ch\,\alpha t + b\,sh\,\alpha t \quad \text{and} \quad Y = a\,sh\,\alpha t + b\,ch\,\alpha t \tag{2.8}$$

are the exact solution of Eqs. (2.6) and (2.7), and they also describe the flow near the critical point with Lagrangian coordinates $a = 0$ and $b = 0$. At first glance, it seems that the flow has nothing to do with the solution of Eq. (2.5). But this is not true. The coordinates of the trajectories of fluid particles for the flow (2.7) satisfy the condition

$$Y^2 - X^2 = b^2 - a^2,$$

and these are also hyperboles, as in the case of Eq. (2.5). Now, the picture of the streamlines is rotated by 45 degrees, so that, not the coordinate axes, but the straight lines $y = \pm x$ serve as asymptotics for hyperbolas. The flow (2.8) can be represented in the following form:

$$X + Y = (a + b)e^{\alpha t}; \quad X - Y = (a - b)e^{-\alpha t}.$$

This means that if we move to the new coordinates of the trajectories of particles $X + Y$ and $X - Y$ and the new Lagrangian variables $a + b$ and $a - b$, then the flow representation (2.8) will have the form (2.5). These transformations correspond to a 45-degree rotation of the coordinate axes, both in the X, Y physical plane and in the plane of Lagrangian coordinates.

2.2.2. *Axisymmetric case*

We now consider the motion of a homogeneous incompressible fluid in cylindrical variables R, Φ, and Z (see Section 1.6.1). Assume that fluid particles move in azimuthal planes ($\Phi = \varphi$), and the radial and axial velocities do not depend on the azimuthal angle ($R_{t\varphi} = 0$ and $Z_{t\varphi} = 0$). The equations of fluid dynamics (1.52) and (1.56) will take the following form:

$$\frac{\partial}{\partial t}(RD) = 0; \quad D = \frac{D(R, Z)}{D(r, z)},$$

$$DR_{tt} = -\frac{1}{\rho}\frac{D(p, Z)}{D(r, z)}; \quad DZ_{tt} = -\frac{1}{\rho}\frac{D(R, p)}{D(r, z)}.$$

The expressions

$$R = re^{\gamma t}, \quad Z = ze^{-2\gamma t}, \quad \text{and} \quad p = p_0 - \frac{\rho\gamma^2}{2}(r^2 e^{2\gamma t} + 4z^2 e^{-4\gamma t}), \tag{2.9}$$

where γ is the constant, are the solutions of this system of equations. The last equation in (2.9) is the Bernoulli integral. The constant p_0 should be chosen so that the pressure remains positive in the considered flow region during the entire observation period. The stream function for the flow (2.9) is equal to $\psi = \gamma R^2 Z$. In each plane $\Phi = \text{const}$, the streamlines will be qualitatively similar to the streamlines shown in Fig. 2.1 The Jacobian $D = \exp(-\gamma t)$ of the transformation from Eulerian to Lagrangian variables does not vanish in the flow region (the condition of one-to-one correspondence of coordinates). The vorticity of the axisymmetric flow is determined

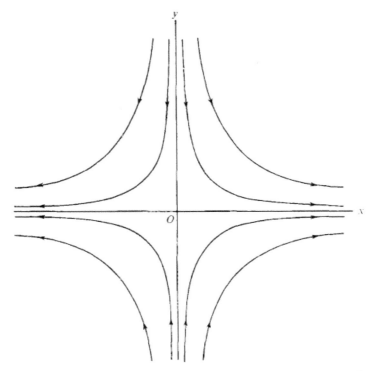

Fig. 2.1. The streamlines of a two-dimensional vortex-free solenoid flow near the critical point, $\psi = \alpha xy$ and $\alpha > 0$ (when $\alpha < 0$, the flow direction reverses).

by the relation

$$\Omega_\Phi = \frac{\partial R_t}{\partial Z} - \frac{\partial Z_t}{\partial R} = -D^{-1} \cdot \left[\frac{D(R_t, R)}{D(r, z)} + \frac{D(Z_t, Z)}{D(r, z)} \right].$$

The vorticity is zero for motion (2.9).

2.3. Flow Initiated by a Point Vortex

Let us continue the review of the simplest exact solutions in Lagrangian variables. Consider an infinitely thin rectilinear vortex tube located in an unbounded fluid. Such an object of fluid dynamics is called a rectilinear vortex line. The fluid motion in all planes perpendicular to the vortex line occurs in the same way; therefore, it is sufficient to consider the flow only, for example, in the XOY plane, and, instead of the vortex line, consider its intersection point with this plane. This point will be referred to as a point vortex. The fluid particles rotate around it in a circle with azimuthal velocity V_{azim} equal to

$$V_{azim} = \frac{\Gamma}{2\pi R_\perp}, \tag{2.10}$$

where $R_\perp$ is the distance to the vortex line and Γ is the vortex intensity. In this case, the positive and negative Γ correspond to the counterclockwise and clockwise circular motion, respectively [4].

In the Lagrangian representation, the flow (2.10) is written as

$$W = X + iY = (a + ib) \exp\left[\frac{i\Gamma}{2\pi(a^2 + b^2)} t \right]. \tag{2.11}$$

We have deliberately written this expression in a complex form since it clearly reflects the physical meaning of the flow. Each fluid particle with initial coordinates a and b rotates along a circle of radius $\sqrt{a^2 + b^2}$ with a frequency equal to $\Gamma/2\pi(a^2 + b^2)$. In the real form, the solution of (2.11) will be rewritten as

$$\begin{aligned} X &= a\cos\frac{i\Gamma}{2\pi(a^2 + b^2)}t - b\sin\frac{i\Gamma}{2\pi(a^2 + b^2)}t; \\ Y &= b\cos\frac{i\Gamma}{2\pi(a^2 + b^2)}t + a\sin\frac{i\Gamma}{2\pi(a^2 + b^2)}t. \end{aligned} \tag{2.12}$$

The correctness of this solution is verified by direct substitution into the system of equations (2.6) and (2.7). There is another way to check this. Denote $\chi = a + ib$, then for the complex conjugate flow velocity $\bar{V} = V_x - iV_y = \overline{W_t}$ (the line is the complex conjugation sign), the following equality is valid:

$$\bar{V} = -\frac{i\Gamma}{2\pi}\frac{1}{W}.$$

For the planar motion of a fluid, the dependence of the complex conjugate velocity on the complex coordinate is the property of the flow potentiality. The incompressibility equation and the non-vorticity condition serve as Cauchy–Riemann conditions for the analytical function $\bar{V}(W)$. Thus, it is proved that the flow (2.12) satisfies the equations of fluid dynamics.

The examples discussed in Sections 2.2 and 2.3 are among those exact solutions when the fluid motion can be described in both Lagrangian and Eulerian representations. This is their distinctive property. In the future, the number of such examples will be extended. Note that, while the point vortex is described relatively simply in the Eulerian description, its Lagrangian representation looks obviously more complicated. It is no longer possible to give a Lagrangian description of the dynamics of two point vortices. In this sense, it should be admitted that it is definitely preferable to study point vortex systems in Eulerian variables.

2.4. Fluid Ellipse Dynamics

In Eulerian coordinates, the description of free-boundary flows has significant difficulties since the boundary condition of constant pressure should be fulfilled on a previously unknown surface. In contrast, in the Lagrangian description, the fluid motion occurs in a given spatial domain of the variables. Let us begin the consideration of free-surface flows with the problem of the evolution of a fluid ellipse neglecting gravity.

Assuming the fluid to be weightless, consider the flows of the following type:

$$X = ae^{\eta(t)}, \quad Y = be^{-\eta(t)}, \quad \text{and} \quad \eta = \int_0^t A(\tau)d\tau. \qquad (2.13)$$

The Lagrangian variables a and b coincide with the initial positions of fluid particles at $t = 0$. Equation (2.13) satisfies continuity equation (2.6). Substitute it into the momentum equation (1.18):

$$-\frac{1}{\rho}p_a = a(A^2 + \dot{A})e^{2\eta}; \quad -\frac{1}{\rho}p_b = b(A^2 - \dot{A})e^{-2\eta}. \tag{2.14}$$

Here, it was taken into account that

$$\dot{\eta} = A \quad \text{and} \quad \ddot{\eta} = \dot{A}. \tag{2.15}$$

Integrating Eq. (2.14), we obtain the expression for pressure:

$$\frac{p}{\rho} = -\frac{1}{2}a^2(\dot{A} + A^2)e^{2\eta} + \frac{1}{2}b^2(\dot{A} - A^2)e^{-2\eta} + f(t). \tag{2.16}$$

Assume that a circle $a^2 + b^2 = r_0^2$ with the center at zero corresponds to the fluid in the plane of Lagrangian variables. At the time $t = 0$, the fluid is concentrated in this circle. In the next moments of time, the circle boundary will start to deform. Since the coordinates of fluid particles, according to Eq. (2.13), satisfy the equality

$$X^2e^{-2\eta} + Y^2e^{2\eta} = r_0^2,$$

then at $t > 0$ an ellipse with semi-axes $r_0 \exp \eta$ and $r_0 \exp(-\eta)$ will be a free surface. At the ellipse boundary, the constant pressure condition should be fulfilled (for simplicity, we assume that it is equal to zero):

$$p|_{a^2+b^2=r_0^2} = 0. \tag{2.17}$$

This is possible if the following relations are valid:

$$(\dot{A} + A^2)e^{2\eta} + (\dot{A} - A^2)e^{-2\eta} = 0; \tag{2.18}$$

$$f(t) = -\frac{r_0^2}{2}(\dot{A} - A^2)e^{-2\eta}.) \tag{2.19}$$

The function f is found from the well-known expression for A and determines the pressure p_0 in the center of the ellipse ($a = b = 0$), which is divided by the density, namely, $f = p_0/\rho$.

Potential Motion 35

Equation (2.18) can be rewritten in terms of $\eta = \eta(\tau)$ [5]:

$$\frac{\ddot{\eta}}{\dot{\eta}} = -\dot{\eta}\, th\, 2\eta. \tag{2.20}$$

Integrating this relation, we obtain

$$\frac{d\eta}{d\tau} = A_0 (ch\, 2\eta)^{-\frac{1}{2}}, \tag{2.21}$$

where $A_0 = A(\tau = 0)$. Let us introduce a new variable $\xi = \exp\eta$ and integrate Eq. (2.21):

$$\sqrt{2}A_0\tau = \int_0^\xi \sqrt{1 + \xi^{-4}}\, d\xi. \tag{2.22}$$

Figure 3 shows diagrams of changes in the horizontal (curve a) and vertical (curve d) axes of the ellipse, normalized to $2r_0$. At the initial time ($\tau = 0$), the axes of the ellipse are the same (the initial shape is a circle). As time increases, the horizontal axis starts to grow and the vertical axis decreases. If one changes the signs in exponents in the expressions for X and Y, the vertical axis will grow. Since $XY = ab$, each particle moves along a hyperbola, as in the flows in Section 2.2 (see Fig. 2.1).

Taylor proposed asymptotic expressions for the dependence (2.22) [5]. For positive η, at $\xi > 1$, the formula has the form

$$\sqrt{2}A_0\tau_+ = C_* + \xi - \frac{1}{2\cdot 3}\xi^{-3} + \frac{1}{8\cdot 7}\xi^{-7}\ldots; \quad C_* = -0.8472,$$

while for negative η at $\xi < 1$, the expansion is written as

$$\sqrt{2}A_0\tau_- = -C_* - \xi^{-1} + \frac{1}{2\cdot 3}\xi^{3} - \frac{1}{8\cdot 7}\xi^{7}\ldots$$

A graphical form of these relations is given in [5] (curves b and c, respectively). The value of the constant C_* was determined by selection. It is clear from this figure that, only within the interval $-0.5 < A_0\tau < 0.5$, the asymptotic expansions differ significantly from the exact solution. For the written asymptotics, the following relation is valid:

$$\tau'_+(\xi) = -\tau'_-\left(\frac{1}{\xi}\right).$$

Since ξ^2 is the ratio of the ellipse semi-axes, it follows from this equality that the ratio of the X-axis to the Y-axis at some point preceding the time $\tau = 0$, when the ellipse becomes a circle, will be the same as the ratio of the Y-axis to the X-axis after the same time since the time $\tau = 0$.

We now find the pressure in the ellipse center. Taking into account Eqs. (2.15), (2.20), and (2.21), we write Eq. (2.19) as follows:

$$\frac{p_0}{\rho} = \frac{2A_0^2 r_0^2}{(\xi^2 + \xi^{-2})^2}.$$

The diagram for the quantity $p_0/\rho A_0^2 r_0^2$ is shown in [5] (curve (e)). The fluid pressure formula (2.16) can be written in the following form:

$$p = \frac{2\rho A_0^2 r_0^2}{(\xi^2 + \xi^{-2})^2}(r_0^2 - a^2 - b^2).$$

The pressure is maximum in the center and drops to zero at the free boundary.

Regardless of another study [5], L.V. Ovsyannikov gave a solution to this problem [6, 7]. The stability of motion (2.13) for two-dimensional potential perturbations in the linear approximation was studied [7, 8]. The flow is stable in a linear approximation if the

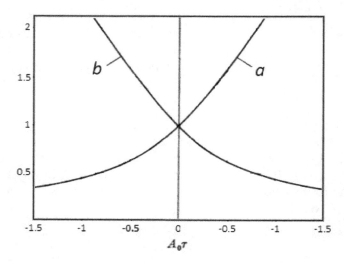

Fig. 2.2. Temporal variation of the horizontal (a) and vertical (b) axes of the ellipse, normalized to $2r_0$ [5].

perturbations of the potential or its derivatives are taken as a stability measure. However, the motion is unstable if deviations of the free boundary from its unperturbed state are considered as a stability measure.

The range of applications of the solution (2.13) can be extended [9]. Let, at the initial time, the free surface have the form of a hyperbola and be set by the equation $a^2 - b^2 = r_*^2$, where r_* is a constant number. Due to Eq. (2.13), the free surface will evolve, obeying the condition $X^2 e^{-2\eta} - Y^2 e^{2\eta} = r_*^2$, i.e., it will remain a hyperbola. The condition of constant pressure on a free surface is written similarly to Eq. (2.18); only the plus sign in the center of the formula should be replaced with a minus sign. For a hyperbolic free surface, the equation for A will have the form

$$(\dot{A} + A^2)e^{2\eta} = (\dot{A} - A^2)e^{-2\eta},$$

where $\dot{A} > A^2$. We differentiate this equality over time and take Eq. (2.15) into account. As a result, we obtain

$$A\ddot{A} - 4\dot{A}^2 + 2A^4 = 0.$$

The first integral of this equation is the expression [9]

$$\frac{\dot{A}^2}{A^8} - \frac{1}{A^4} = Q > 0,$$

where Q is a positive constant (due to the fact that $\dot{A} > A^2$), so that

$$\dot{A}^2 = A^4(1 + QA^4),$$

and finally

$$t = \pm \int \frac{dA}{A^2(1 + QA^4)^{\frac{1}{2}}}.$$

All possible flow patterns of the type considered have been studied in detail [9]. The fluid can be located both between the branches of hyperbolas and inside each branch. Over the course of time, the branches of hyperbolas (free boundaries of the flow) deform so that the angle between their asymptotics can both increase and decrease, depending on the sign of A (which determines the velocity direction). The angle between the asymptotics and the X-axis is equal to

$$\alpha = arctg \left(\frac{\dot{A} + A^2}{\dot{A} - A^2} \right)^{\frac{1}{2}}.$$

When the fluid moves, the angle remains either greater than $45°$ ($\dot{A} > 0$) or less than this value ($\dot{A} < 0$). As a mark of respect to Dirichlet, who proposed to use velocity fields linearly dependent on spatial coordinates to study the motion of fluid volumes with a free boundary [10, 11], Longuet-Higgins called his flows Dirichlet hyperbolas [12].

As Taylor experimentally proved [13], the angle at the top of the crest of a two-dimensional gravity standing wave of the maximum amplitude is close to $90°$. For Longuet-Higgins' solutions, this value of the angle between the asymptotics is critical. Therefore, it can be argued that flows with a free surface in the form of hyperbolas model the local properties of the limiting standing wave [9, 12].

Finally, an interesting special case of flows (2.13) occurs if $A = \pm 1/t$ [9]. Let us choose without loss of generality that $A = t^{-1} > 0$ ($t > t_0 > 0$). Then, from Eq. (2.15) we have

$$\frac{p}{\rho} = -\frac{b^2}{t^4} + f(t).$$

Assume that $f(t) = h^2/t^4$. Then, the values of $b = \pm h$ correspond to zero-pressure surfaces. In the physical plane, these will be two horizontal planes parallel to the X-axis, spaced by a distance b/t^2 from it and approaching at a speed of $2b/t^3$. The case $A = -t^{-1}$ is considered in a similar way, but now the flow is concentrated between two planes parallel to the Y-axis.

2.5.   Movement of a Fluid Cone with a Free Side Surface

Let us consider, following a study [14], a fluid volume having the shape of a circular cone. Let its vertex be at the origin of coordinates, the Z-axis coincide with the axis of the cone and be directed vertically upward, and the vertex angle be equal to 2Θ. The lateral surface of the cone is free, and the pressure constancy condition should be

Potential Motion

fulfilled on it (assume that it is equal to zero). The fluid is not limited vertically. The initial flow area is specified by the condition

$$(a^2 + b^2)l^2 - c^2 \leq 0; \quad 0 \leq c \leq \infty, \tag{2.23}$$

where l is a constant value. The angle between the cone generatrix and the Z-axis at the initial time is $\Theta_0 = \arctan(1/l)$.

We will seek a solution in the form

$$X = a \cdot m(t); \quad Y = b \cdot m(t); \quad Z = c \cdot m^{-2}(t), \tag{2.24}$$

where $m(t)$ is an unknown function. The coordinates a, b, and c are identified with the initial position of the fluid particles, so that

$$m(0) = 1. \tag{2.25}$$

Equation (2.24) is equivalent to the following Eulerian representation:

$$V_x = m'm^{-1}X; \quad V_Y = m'm^{-1}Y; \quad V_Z = -2m'm^{-1}Z.$$

The vorticity of this spatial flow is obviously zero.

Equation (2.24) satisfies continuity equation (1.11). In this case, $D_0 = 1$. Gravity will be neglected. After substituting Eq. (2.24) into the momentum equation (1.16), we have

$$amm_{tt} = -\frac{1}{\rho}p_a; \quad bmm_{tt} = -\frac{1}{\rho}p_b; \quad cm^{-2}(m^{-2})_{tt} = -\frac{1}{\rho}p_c.$$

Integrating these equations, we obtain the expression for pressure:

$$-\frac{p}{\rho} = \frac{1}{2}[(a^2 - b^2)mm_{tt} + c^2m^{-2}(m^{-2})_{tt}] + f(t).$$

The pressure on the surface (2.23) will remain zero if the following conditions are fulfilled:

$$\frac{mm_{tt}}{l^2} = -m^{-2}(m^{-2})_{tt}; \quad f = 0.$$

The first of them is reduced to the following equation:

$$(m^6 - 2l^2)m_{tt} + \frac{6l^2m_t^2}{m} = 0,$$

which has the first integral

$$m_t^2 = M \frac{m^6}{m^6 - 2l^2}; \quad M = \text{const.} \tag{2.26}$$

When choosing the value of M, it should be taken into account that when $t = 0$, the denominator turns to zero if $l^2 = 1/2$. Therefore,

$$M = \begin{cases} m_0^2(1 - 2l^2); \ m_0^2 = m_t^2(0) & \text{if } l^2 \neq 1/2 \\ 0 & \text{if } l^2 = 1/2 \end{cases} . \tag{2.27}$$

If $l^2 = 1/2$, then $m_t = 0$ and $m(t) \equiv 1$, i.e., a straight circular cone with an initial vertex angle equal to $2\Theta_* = 2\arctan\sqrt{2} \approx 109.46°$, can only be at rest.

Let us analyze two cases sequentially.

(a) $l > 2^{-\frac{1}{2}}$ and $\Theta_0 < \Theta_*$. If $m_0 > 0$, then $m(t)$ increases, and it follows from Eqs. (2.26) and (2.27) that

$$\int_1^m \frac{\sqrt{2l^2 - m^6}}{m^3} dm = m_0 t \sqrt{2l^2 - 1}.$$

Thus, $m \to m_* = (2l^2)^{1/6}$ at $t \to t_*$, where t_* is determined by the formula

$$t_* = \frac{1}{m_0\sqrt{2l^2 - 1}} \cdot \int_1^{m_*} m^{-3}\sqrt{2l^2 - m^6}\,dm.$$

According to Eqs. (2.4) and (2.23), the free surface changes according to the law

$$\frac{X^2 + Y^2}{m^2 l^{-2}} - \frac{Z^2}{m^{-4}} = 0,$$

i.e., it remains a cone with an angle at the vertex

$$2\Theta(t) = 2\arctan\left(m^3/l\right). \tag{2.28}$$

The angle $\Theta \to \Theta_*$ at $t \to t_*$.

At $m_0 < 0$, the function $m(t)$ decreases, and $m \to 0$ when $t \to \infty$, so that $\Theta \to 0$. The cone breaks over an infinite time.

Potential Motion   41

(b) $l < 2^{-\frac{1}{2}}$ and $\Theta_0 > \Theta_*$. At $m_0 > 0$, the function $m(t)$ increases, and $m \to \infty$, following the law

$$\int_1^m m^{-3}\sqrt{m^6 - 2l^2}\, dm = m_0 t\sqrt{1 - 2l^2}.$$

In this case, $\Theta(t) \to \pi/2$ (see Eq. (2.28)) and the cone tends to merge with the plane $Z = 0$.

If $m_0 < 0$, then $m(t)$ decreases and reaches its minimum m_* over a time

$$t_* = \frac{1}{|m_0|\sqrt{1 - 2l^2}} \cdot \int_{m_*}^1 m^{-3}\sqrt{m^6 - m_*^6}\, dm.$$

The pressure inside the fluid is determined by the expression

$$p = \pm\frac{3\rho m_0^2(1 - 2l^2)m^6}{(m^6 - 2l^2)^2}[c^2 - l^2(a^2 + b^2)],$$

where the function $m(t)$ satisfies Eq. (2.26). The plus sign in this expression corresponds to the case a and the minus sign to the case b.

The limiting angle Θ_* should be separately discussed. Mack pointed out that an axially symmetric standing gravity wave has a cutoff internal angle of $2\arctan\sqrt{2}$ [15]. Based on this result, it can be assumed that the considered motion of the fluid cone models the fluid dynamics in the vicinity of the crest of the standing wave of the maximum amplitude.

2.6. Spherical Bubble Dynamics

Consider the evolution of a spherical cavity created by an underwater explosion. Since the acceleration in the radial direction is much greater than the acceleration of gravity g, the fluid can be considered weightless in a first approximation. We also assume that the velocities are low compared to the speed of sound so that the fluid is incompressible. It is required to determine the dynamics of the bubble boundary, provided that the motion is purely radial and the pressure dependence at the bubble boundary is known and equal to $p_b(t)$. This problem can be solved gracefully in Eulerian variables [3].

42   *Analytical Fluid Dynamics in Lagrangian Variables*

We assume that the azimuthal and meridional components of the velocity are equal to zero and the radial velocity is determined by the relation

$$V_R = \frac{R_b^2 \dot{R}_b}{R^2}, \tag{2.29}$$

where R is the distance measured from the center of the bubble and the function $R_b(t)$ describes the motion of its boundary. The velocity field (2.29) satisfies the continuity equation and corresponds to a vortex-free flow with the potential equal to

$$\varphi(R, t)(m^{-2}) = -\int_R^\infty V_r dR = -\frac{R_b^2 \dot{R}_b}{R}.$$

Using the expression for the Cauchy–Lagrange integral, we find the fluid pressure

$$\frac{p(R, t) - p_0}{\rho} = -\frac{\partial \varphi}{\partial t} - \frac{V_R^2}{2} = \frac{2R_b \dot{R}_b^2 + R_b^2 \ddot{R}_b}{R} - \frac{1}{2} \frac{R_b^4 \dot{R}_b^2}{R^4}, \tag{2.30}$$

where p_0 is a constant pressure away from the bubble. If, according to the explosion results, the pressure inside the bubble is known, i.e.,

$$p(R, t)(m^{-2})\big|_{R=R_b} = p_b(t),$$

then the radius of the bubble satisfies the differential equation

$$R_b \ddot{R}_b + \frac{3}{2} \dot{R}_b^2 = \frac{p_b - p_0}{\rho}. \tag{2.31}$$

This is an ordinary differential equation that should be solved numerically.

As a gas bubble expands and causes the surrounding water to move radially, the pressure in the gas decreases approximately adiabatically, being inversely proportional to the bubble radius of a high degree. At later stages of expansion, the pressure in the bubble drops sharply below p_0 and it can be considered empty ($p_b \ll p_0$). The external pressure p_0 is constant, and Eq. (2.31) can be integrated once [3]. Thus, we find

$$R_b^3 \dot{R}_b^2 = \frac{2}{3} \frac{p_0}{\rho} (R_{bm}^3 - R_b^3),$$

where R_{bm} is the maximum radius of a bubble as it stops expanding and starts to shrink. Further, the relation between the time t and the

Potential Motion 43

radius R_b of an effectively empty bubble in a fluid, in which a constant uniform pressure $p_0 > 0$ is maintained at infinity, is determined by the quadrature

$$|t - t_m| = \sqrt{\frac{3\rho}{p_0}} \int_{R_b}^{R_{bm}} \frac{dR_b}{\sqrt{\left(\frac{R_{bm}}{R_b}\right)^3 - 1}}, \tag{2.32}$$

where t_m is the time at which $R_b = R_{bm}$. This formula is valid for both positive and negative values of the difference $t - t_m$. Herein, the motion in the contraction phase $t > t_m$ is a simple reversal of the bubble motion in the expansion phase.

Equations (2.31) and (2.32) describe the bubble boundary dynamics. The Lagrangian representation of the considered flow is found as a result of integrating Eq. (2.29):

$$R = \left(r^3 + 3 \int_0^t R_b^2 \dot{R}_b dt\right)^{1/3}, \tag{2.33}$$

where $r = R|_{t=0}$ is the Lagrangian coordinate. Equation (2.33) reflects a rare case where the exact solution for a spatial flow can be written in Lagrangian coordinates.

Strictly speaking, to solve this problem in spherical Lagrangian coordinates, it should be noted that the fluid particles change only their radial coordinate (see Section 1.6.2) according to the law (2.33). This expression satisfies continuity Eq. (1.58). In this approach, the formula for pressure can be obtained from Eq. (1.59). However, substituting Eq. (2.33) into Eq. (2.30) indicates how more cumbersome the purely Lagrangian approach looks for this problem.

2.7. John's Method

In the Lagrangian description of two-dimensional flows, a free surface can be correlated with a parametrically specified curve, where one of the Lagrangian coordinates serves as a parameter. This mathematical technique makes it possible to advance in solving a number of nonlinear problems [16]. Consider the two-dimensional motion of an incompressible, vortex-free fluid with a free surface. This motion is described by the complex potential $\phi(z, t) = \varphi(x, y, t) + i\psi(x, y, t)$, where $\phi(z,t)$ is an analytical function of the complex variable $z = x + iy$ and time. The fluid velocity is determined by the relation

$$\frac{dz}{dt} = \overline{\phi_z}. \tag{2.34}$$

Let us denote the boundary of the free surface by C. This free surface consists of the same particles, and the pressure on it is constant.

Let χ be the Lagrangian coordinate for particles on line C. Then, this curve can be represented parametrically as

$$z = f(\chi, t). \tag{2.35}$$

Constant pressure at a free boundary is equivalent to the fact that the acceleration

$$\frac{d^2 z}{dt^2} + ig$$

of the free-surface particles is perpendicular to the line C (g is the free-fall acceleration and the y-axis is directed vertically upward). Since the derivative f_χ is tangential to C, this condition can be rewritten in the form

$$f_{tt} + ig = ir(\chi, t)f_\chi, \tag{2.36}$$

where $r(\chi, t)$ is a real-valued function and irf_χ indicates the pressure reduction direction. Thus, the most general form of a two-dimensional free surface can be represented by Eq. (2.35), where the function $f(\chi, t)$ for real values of χ is the solution of Eq. (2.36), which is a partial differential equation of a parabolic type. Determining the appropriate type of the function $r(\chi, t)$ is the main difficulty of the general problem. If the function $r(\chi, t)$ is given, the problem is reduced to solving Eq. (2.36) under the necessary boundary conditions. Note also that for a given function $r(\chi, t)$, this equation is a linear equation with respect to $f(\chi, t)$, and therefore its solution can be obtained using the superposition method.

In order to find a fluid flow with a free surface C, one should calculate the function $\phi(z, t)$, which is analytical in the flow region and satisfies the boundary condition

$$f_t = \overline{\phi_z}$$

Potential Motion 45

on the curve C, in accordance with Eq. (2.34). Consider ϕ as a function of χ and t. Then,

$$\phi_\chi(\chi, t) = \phi_z z_\chi = \overline{f_t(\chi, t)} f_\chi(\chi, t),$$

or, given that χ is real valued on C, we have

$$\phi_\chi(\chi, t) = \overline{f_t(\bar\chi, t)} f_\chi(\chi, t). \qquad (2.37)$$

If we assume that the line C is described by an analytical function χ and the function f_t is also analytical, then the right-hand side of Eq. (2.37) is an analytical function. Assuming the variable χ in Eq. (2.37) to be complex, we obtain an analytical function ϕ for χ and hence for z. The function will represent the flow potential with a free surface C specified by the function $f(\chi, t)$. Consequently, any analytical function f of Eq. (2.36) with coefficient r, which remains real valued for real χ, determines some flow with a free boundary, with the potential $\phi(\chi, t)$ calculated by integration (2.37). We emphasize that in this case χ coincides with the Lagrangian variable on the free boundary, but it is just a parameter inside the fluid. The functions ϕ and z are specified parametrically through this parameter, determining the form of $\phi(z, t)$. Due to this, the method is sometimes also called semi-Lagrangian.

2.7.1. *Exact solution for a vortex-free wave*

In general, John's method is applicable to describe non-stationary potential flows, but we assume that the motion is steady. Let z_0 denote the point on the free surface. We now determine the Lagrangian coordinate χ of a particle located on the surface from the condition that $-\chi$ is the time at which the particle occupies the position z_0. Moving from the position z_0 to z requires the same time β for all particles in a steady flow. Therefore, the function $z = f(\chi, \beta - \chi)$ should not depend on χ for each β. Hence,

$$z = f(\chi, t) = z(\beta) \quad \text{and} \quad \beta = \chi + t. \qquad (2.38)$$

Then, it follows from Eq. (2.36) that the quantity $r(\chi, t) = (f_{tt} + ig)/if_\chi$ should be a function of β only, e.g.,

$$r(\chi, t) = s'(\beta),$$

and Eq. (2.36) is reduced to an ordinary differential equation of the form

$$z''(\beta) + ig = is'(\beta)z'(\beta), \tag{2.39}$$

where $s'(\beta)$ and therefore $s(\beta)$ are real-valued functions for real β. It follows from Eq. (2.37) that the complex potential ϕ is also a function of β and is determined using the quadrature from the equation

$$\frac{d\phi}{d\beta} = \overline{z'(\bar{\beta})}z'(\beta). \tag{2.40}$$

Here, $\beta = \chi + t$ should be considered complex.

Let $s(\beta) = \omega\beta$, where ω is a constant with frequency dimension. Solving Eq. (2.39), we obtain

$$z(\beta) = B + Ae^{i\omega\beta} + \frac{g}{\omega}\beta,$$

where, without loss of generality, we can assume $B = 0$, and consider A to be real and positive, so that finally

$$z(\beta) = \frac{g}{\omega}\beta + Ae^{i\omega\beta}. \tag{2.41}$$

The free surface of the flow is a trochoid with amplitude A and period $\lambda = 2\pi g/\omega^2$. The expression for $\omega^2 = 2\pi g/\lambda$ coincides with the dispersion relation of gravity linear waves in deep water.

Following Eq. (2.40), for the potential, we obtain

$$\frac{d\phi}{d\beta} = \left(\frac{g}{\omega} - iA\omega e^{-i\omega\beta}\right)\left(\frac{g}{\omega} + iA\omega e^{i\omega\beta}\right);$$

hence,

$$\phi = \left(\frac{g^2}{\omega^2} + A^2\omega^2\right)\beta + \frac{2Ag}{\omega}\cos\omega\beta. \tag{2.42}$$

The velocity $d\phi/dz$ becomes infinite if $dz/d\beta = 0$. This yields

$$\omega\beta = 2\pi\left(n + \frac{1}{4}\right) - i\ln\frac{\lambda}{2\pi A},$$

where n is an integer. The corresponding z values are singular points:

$$z = \lambda\left\{\left(n + \frac{1}{4}\right) + \frac{i}{2\pi}\left(1 - \ln\frac{\lambda}{2\pi A}\right)\right\}. \tag{2.43}$$

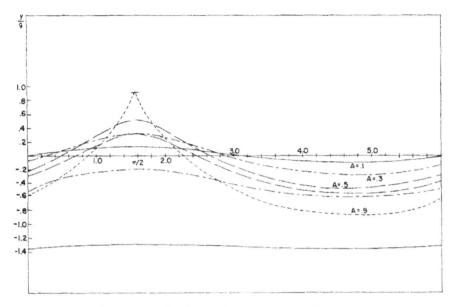

Fig. 2.3. Stationary flow over a wavy bottom.

Such points should be excluded from the flow. For this purpose, the bottom one can take the streamline $\operatorname{Im}\phi = \mathrm{const}$, which passes through the branching points or above the points specified by Eq. (2.43).

Figure 2.3, taken from another study [16], shows a free surface and a bottom surface formed by a streamline passing through singular points for various values of A/λ. The measurement units in the diagram are chosen so that $\omega = g = 1$ and $\lambda = 2\pi$. For a small value of A/λ, the fluid depth is of the order of $\lambda/\ln(\lambda/A)$, which is significantly greater than A. The elevation of the bottom surface is of the order of A^2/λ and is much smaller than the amplitude A of the free surface. On the other hand, when the ratio A/λ is close to $1/2\pi$, the fluid depth is small and the bottom surface resembles a free boundary. However, the amplitude A must remain strictly less than $\lambda/2\pi$, i.e., the trochoidal profile of the free surface has no sharpening points.

For the motion given by Eq. (2.41), the free surface remains unchanged and fixed. Each fluid particle has a horizontal velocity varying between the values $g/\omega - \omega A$ and $g/\omega + \omega A$. In a coordinate system moving at a constant velocity g/ω, the solution will have the form of a traveling wave. This motion is analytically described as follows:

$$z = \frac{g}{\omega}\chi + A\exp\{i\omega(\chi + t)\}.$$

The form of the complex potential in a moving reference frame corresponds to the transformation $\phi \to \phi - gz/\omega$ and is equal to

$$\phi = \omega^2 A^2(\chi + t) + \frac{Ag}{\omega}\exp\{-i\omega(\chi + t)\}.$$

The complex potential ϕ is a function of $z + \frac{g}{\omega}t$, so that the wave propagates to the left with velocity g/ω. The bottom surface moves in a similar way. However, for $A/\lambda \ll 1$, it can be chosen so deep that the fluid motion will be reduced to small fluctuations of an infinitely deep fluid.

2.7.2. John's parabola

We now choose the function z in the form

$$z = -it\chi^2 + \alpha(t)\chi + i\beta(t), \tag{2.44}$$

where α and β are real values. For real χ (free surface), this formula is a parametric representation of a parabola:

$$y = -t\chi^2 + \beta; \quad x = \alpha\chi.$$

Substituting Eq. (2.44) into Eq. (2.36), we obtain restrictions on the choice of time-dependent functions:

$$\alpha\alpha'' = 2t(\beta'' + g); \quad r = \frac{\alpha''}{2t}. \tag{2.45}$$

The fluid velocity is determined by the relation

$$\phi_z = \frac{\overline{dz}}{dt} = \overline{f_t(\bar{\chi}, t)} = i\chi^2 + \alpha'\chi - i\beta'. \tag{2.46}$$

Equations (2.44) and (2.45) determine a parametrically specified flow with a free surface in the form of a parabola. It is worth considering

Potential Motion

that the function $\chi(z)$ has a branching point $\chi_* = -\frac{i\alpha}{2t}$. To make the velocity field unambigious, $\phi_z = \phi_\chi / Z_\chi$, one should set

$$\alpha' = -\frac{\alpha}{t}.$$

Using this relation, we find formulas for time-varying functions,

$$\alpha = \frac{C}{t}, \quad r = \frac{C}{t^4}, \quad \text{and} \quad \beta = \frac{C^2 t^{-3}}{12} - \frac{gt^2}{2},$$

where C is the constant, and the expression for the potential

$$\phi(z) = -\frac{z^2}{2t} + i\left(\frac{C^2}{3t^4} + \frac{gt}{2}\right)z.$$

The free surface changes with time, remaining a parabola:

$$y = -\frac{t^2 x^2}{C^2} + \frac{C^2}{12t^3} - \frac{gt^2}{2}.$$

The transition to a free-falling system allows the last term to be discarded.

2.7.3. Ascending flow

Let $r = 2/t$. Equation (2.36) takes the form

$$z_{tt} + ig = \frac{2i}{t}z_\chi.$$

Its solution can be written as follows:

$$z = \frac{1}{2}gt\chi. \tag{2.47}$$

Due to this relation, we determine the flow potential derivative

$$\phi_\chi(\chi, t) = \overline{z_t(\bar{\chi}, t)}z_\chi(\chi, t) = \frac{1}{4}g^2\chi t$$

and then the potential itself

$$\phi(\chi, t) = \frac{1}{8}g^2\chi^2 t. \tag{2.48}$$

In Eulerian variables, the flow given by Eqs. (2.47) and (2.48) is described by the potential $\phi = z^2/2t$. The equation of streamlines has

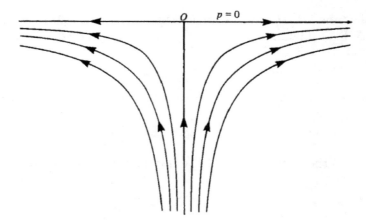

Fig. 2.4. Ascending flow (snapshot).

the form $\psi = xy/t = \text{const}$ (see Fig. 2.4). The velocity components are

$$V_x = \frac{x}{t} \quad \text{and} \quad V_y = -\frac{y}{t},$$

respectively.

From the momentum equation on the X-axis,

$$-\frac{p_x}{\rho} = \frac{\partial V_x}{\partial t} + V_x \frac{\partial V_x}{\partial x} + V_y \frac{\partial V_x}{\partial y},$$

it follows that $p_x = 0$. This confirms the fact that the pressure on the free surface is constant.

*   *   *

In a series of papers [17–22], Longuet-Higgins considered a number of new applications of John's parabola and proposed a more general way for parametric representation of the potential flow when the parameter being used no longer correlates with the Lagrangian coordinate [18]. On the free boundary, in addition to the constant pressure condition ($p = 0$), the condition $dp/dt = 0$ (similar to the kinematic condition) is imposed. For a weightless fluid, Longuet-Higgins (employing numerical calculations) studied various modes of surface wave breaking [20–22]. New [23] and Greenhow [24] employed John's parabola in the numerical simulation of a breaking wave.

References

[1] Loitsyanskii, L.G. (1966). *Mechanics of Liquids and Gases* (Pergamon Press, Oxford, New York).

[2] Ovsyannikov, L.V. (1967). *General Equations and Examples. The Problem of Unsteady Fluid Motion with a Free Boundary* [in Russian], (Nauka, Novosibirsk).

[3] Batchelor, G.K. (1967). *An Introduction to Fluid Dynamics* (Cambridge University Press, Cambridge).

[4] Kochin, N.E., Kibel, I.A. and Roze, N.V. (1964). *Theoretical Hydromechanics*, Vol. 1 (Interscience Publ., New York).

[5] Taylor, G. (1960). Formation of thin flat sheets of water. *Proc. R. Soc. Lond.*, 259(1296), pp. 1–17.

[6] Ovsyannikov, L.V. (1965). On one class of unsteady motion of an incompressible fluid. *Proc. 5th Session of the Scientific Council on the National Economic Use of the Explosion* [in Russian], (Ilim, Frunze).

[7] Nalimov, V.I. and Pukhnachev, V.V. (1975). *Unsteady Motions of an Ideal Fluid with a Free Boundary* [in Russian], (Publishing House of NSU, Novosibirsk).

[8] Pukhnachev, V.V. (1971). Small perturbations of plane unsteady motion of an ideal incompressible fluid with a free boundary in the shape of an ellipse. *J. Appl. Mech. Tech. Phys.*, 12(44), pp. 530–538.

[9] Longuet-Higgins, M S. (1972). A class of exact, time-dependent, free-surface flows. *J. Fluid Mech*, 55(3), pp. 529–543.

[10] Dirichlet, P.L. (1860). Untersuchungen überein problem der hydrodynamik. *Abh. Kön. Ges. Wiss. Göttingen*, 8, pp. 3–42.

[11] Lamb, H. (1932). *Hydrodynamics*, 6th edn. (Cambridge University Press, Cambridge).

[12] Longuet-Higgins, M.S. (1983). Bubbles, breaking waves and hyperbolic jets at a free surface. *J. Fluid Mech.*, 127, pp. 103–121.

[13] Taylor, G.I. (1953). An experimental study of standing waves. *Proc. R. Soc. London*, 218(1132), pp. 44–59.

[14] Andreev, V.K. (1992). *Stability of the Unsteady Motion of a Fluid with a Free Boundary* [in Russian], (Nauka, Novosibirsk).

[15] Mack, L.R. (1962). Periodic, finite-amplitude, axisymmetric gravity waves. *J. Geophys. Res.*, 67, pp. 829–843.

[16] John, F. (1953). Two-dimensional potential flows with a free boundary. *Comm. Pure Appl. Math.*, 6, pp. 497–503.

[17] Longuet-Higgins, M.S. (1976). Self-similar, time-dependent flows with a free surface. *J. Fluid Mech.*, 73(4), pp. 603–620.

[18] Longuet-Higgins, M.S. (1980). A technique for time-dependent free-surface flows. *Proc. Roy. Soc. London*, 371, pp. 441–451.

[19] Longuet-Higgins, M.S. (1980). On the forming of sharp corners at a free surface. *Proc. Roy. Soc. London*, 371, pp. 453–478.
[20] Longuet-Higgins, M.S. (1981). On the overturning of gravity waves. *Proc. Roy. Soc. London*, 376(1766), pp. 377–400.
[21] Longuet-Higgins, M.S. (1982). Parametric solutions for breaking waves. *J. Fluid Mech.*, 121, pp. 403–424.
[22] Longuet-Higgins, M.S. (1983). Rotating hyperbolic flow: particle trajectories and parametric representation. *Quart. J. Appl. Math.*, 36(2), pp. 247–270.
[23] New, A.L. (1983). A class of elliptical free-surface flows. *J. Fluid Mech.*, 130, pp. 219–239.
[24] Greenhow, M. (1983). Free-surface flows related to breaking waves. *J. Fluid Mech.*, 134, pp. 259–275.

Chapter 3

Vortex Motion

"There are many reasons why it is more convenient
to represent fluid motion in terms of vorticity, not
speed, despite the simpler physical meaning of the
latter."

— J. Batchelor, *An Introduction to Fluid Dynamics* [1]

3.1. Complex form of Fluid Dynamics Equations

Complex variables are convenient to use in the Eulerian approach to describe potential flows, but they are no longer effective for studying vortex flows. At the same time, in Lagrangian analysis, the transition to complex coordinates in a number of cases contributes to an effective analytical description of vortices.

Let us turn in the Lagrange equations to Lagrangian variables

$$\chi = a + ib, \quad \bar{\chi} = a - ib, \quad c = c \tag{3.1}$$

and complex coordinates of the trajectory of fluid particles

$$W = X + iY, \quad \bar{W} = X - iY, \quad Z = Z. \tag{3.2}$$

Hereafter, the bar denotes a complex conjugate quantity. The continuity equation in the new variables can be represented as follows

54   *Analytical Fluid Dynamics in Lagrangian Variables*

(see Eq. (1.11)):

$$\frac{\partial}{\partial t} \begin{vmatrix} W_\chi & W_{\bar\chi} & W_c \\ \bar W_\chi & \bar W_{\bar\chi} & \bar W_c \\ Z_\chi & Z_{\bar\chi} & Z_c \end{vmatrix} = 0. \tag{3.3}$$

In these variables, (3.1) and (3.2), the momentum equation (1.18) will be rewritten as follows:

$$\frac{1}{2}(W_{tt}\bar W_\chi + \bar W_{tt} W_\chi) + Z_{tt} Z_\chi = -\frac{p_\chi}{\rho} - H_\chi; \tag{3.4}$$

$$\frac{1}{2}(W_{tt}\bar W_{\bar\chi} + \bar W_{tt} W_{\bar\chi}) + Z_{tt} Z_{\bar\chi} = -\frac{p_{\bar\chi}}{\rho} - H_{\bar\chi}; \tag{3.5}$$

$$\frac{1}{2}(W_{tt}\bar W_c + \bar W_{tt} W_c) + Z_{tt} Z_c = -\frac{p_c}{\rho} - H_c. \tag{3.6}$$

The appearance of the factor $\frac{1}{2}$ is related to the type of representation of (3.1) and (3.2). It can be eliminated if the complex coordinates of the particle trajectory are "compressed" by $\sqrt{2}$ times.

3.2.   Jacobian form of Equations for Two-Dimensional Flows

The equations of fluid dynamics describing the two-dimensional motion of an ideal incompressible fluid are obtained from systems (3.3)–(3.6), if we assume $Z = c$ and $\frac{\partial}{\partial c} = 0$, and have the form

$$\frac{\partial}{\partial t} \begin{vmatrix} W_\chi & W_{\bar\chi} \\ \bar W_\chi & \bar W_{\bar\chi} \end{vmatrix} = 0; \tag{3.7}$$

$$\frac{1}{2}(W_{tt}\bar W_\chi + \bar W_{tt} W_\chi) = -\frac{p_\chi}{\rho} - H_\chi; \tag{3.8}$$

$$\frac{1}{2}(W_{tt}\bar W_{\bar\chi} + \bar W_{tt} W_{\bar\chi}) = -\frac{p_{\bar\chi}}{\rho} - H_{\bar\chi}. \tag{3.9}$$

We now exclude the pressure and potential of external forces from these equations. To do this, we take the derivative with respect to $\bar\chi$

from Eq. (3.8), the derivative with respect to χ from Eq. (3.9), and subtract the second from the first one. As a result, we obtain

$$\frac{\partial}{\partial t}(W_{t\chi}\bar{W}_{\bar{\chi}} - W_{t\bar{\chi}}\bar{W}_{\chi} + \bar{W}_{t\chi}W_{\bar{\chi}} - \bar{W}_{t\bar{\chi}}W_{\chi}) = 0.$$

This equation can be reduced. We add it up with the time-differentiated equation (3.7). As a result, we arrive at the equation

$$\frac{\partial}{\partial t}(W_{t\chi}\bar{W}_{\bar{\chi}} - W_{t\bar{\chi}}\bar{W}_{\chi}) = 0. \tag{3.10}$$

Thus, the system of equations of two-dimensional fluid dynamics can be written in the form of two conservation laws [2, 3]:

$$\frac{D(W, \bar{W})}{D(\chi, \bar{\chi})} = \frac{D(W_0, \bar{W}_0)}{D(\chi, \bar{\chi})} = D_0(\chi, \bar{\chi}); \tag{3.11}$$

$$\frac{D(W_t, \bar{W})}{D(\chi, \bar{\chi})} = \frac{D(W_{t0}, \bar{W}_0)}{D(\chi, \bar{\chi})} = i\frac{\Omega}{2}D_0, \tag{3.12}$$

where W_0 is the complex coordinate of the particle trajectory at the initial time and the time-independent function $\Omega(\chi, \bar{\chi})$ is the vorticity. A distinctive feature of this form of recording is its compactness.

The representation of flows in complex Lagrangian coordinates, as will be shown, turns out to be convenient for solving a number of problems of vortex dynamics and for describing waves on water.

3.3. The Simplest Types of Vortex Flows

We start our acquaintance with the Lagrangian description of specific vortex flows with the simplest types of particle motion, namely, translational, circular, elliptical, and helical.

3.3.1. *Inertial flows*

Assume that no external forces act on the fluid and the pressure inside it is the same everywhere. Then, the momentum equation (1.18) is automatically satisfied if the acceleration of each particle

56 *Analytical Fluid Dynamics in Lagrangian Variables*

is zero:

$$X_{tt} = 0; \quad Y_{tt} = 0; \quad Z_{tt} = 0. \tag{3.13}$$

It is natural to call such flows inertial. Solutions of Eq. (3.13) must also satisfy continuity equation (1.11). It is easy to verify that both conditions are fulfilled for the expressions

$$X = a + U(b, c)t, \quad Y = b, \quad \text{and} \quad Z = c, \tag{3.14}$$

where $U(b, c)$ is an arbitrary function. Equation (3.14) describes the non-uniform drift of fluid particles in the direction of the x-axis. In the special case where the function U depends only on one variable, e.g., b, Eq. (3.14) describes a two-dimensional shear flow.

The vorticity of the flow (3.14) is equal to (see Eq. (1.31))

$$\Omega_x = 0, \quad \Omega_y = U'_c, \quad \Omega_z = -U'_b. $$

3.3.2. *Flows with circular streamlines*

The simplest example of this kind of flow is the uniform fluid rotation. It is convenient to write its representation in complex coordinates:

$$W = X + iY = (a + ib)\exp(i\omega t) = \chi \exp(i\omega t). \tag{3.15}$$

In this representation, each fluid particle with the coordinate $\chi = a + ib$ moves along a circle of radius $|\chi|$ with a constant angular velocity ω. Equation (3.15) satisfies the system of Eqs. (3.11) and (3.12), and the vorticity of the flow Ω is equal to twice the value of the angular velocity ($\Omega = 2\omega$).

In real Lagrangian variables, uniform rotation of the fluid is given by the formulas

$$X = a \cos \omega t - b \sin \omega t,$$

$$Y = a \sin \omega t + b \cos \omega t.$$

This system is sometimes written using a rotation matrix,

$$\vec{X} = \begin{pmatrix} \cos \omega t & -\sin \omega t \\ \sin \omega t & \cos \omega t \end{pmatrix} \vec{a}$$

where $\vec{X}$ and $\vec{a}$ are Eulerian and Lagrangian two-dimensional vectors, respectively.

Equation (3.15) is easily generalized to the case of non-uniform rotation of the fluid, which is described by the relation

$$W = \chi \exp[i\omega_*(|\chi|^2)t], \qquad (3.16)$$

where ω_* is some arbitrary function. Now, fluid particles located at different distances from the origin of coordinates rotate around it with different angular velocities determined by the form of the function $\omega_*(|\chi|^2)$. The flow vorticity (3.16) is obtained from Eqs. (3.11) and (3.12) and is equal to

$$\Omega = 2\left(\omega_* + |\chi|^2 \frac{\partial \omega_*}{\partial |\chi|^2}\right).$$

In the case of uniform rotation, the second term in the parentheses turns to zero. If ω_* decreases inversely proportional to $|\chi|^2$, then the flow is potential. This case corresponds to the fluid motion outside the circular vortex. In fluid dynamics, a model combining these two flows is popular (the Rankine circular vortex [4]).

3.3.3. *Flow inside the ellipse*

Consider a flow given by the expression

$$W = \frac{1}{2}(A + B)\chi \exp(i\gamma t) + \frac{1}{2}(A - B)\bar{\chi} \exp(-i\gamma t). \qquad (3.17)$$

Here, A and B are constants (the reason for their introduction will become clear later). It can be verified by direct substitution that Eq. (3.17) satisfies the system of Eqs. (3.11) and (3.12). The motion of a fluid particle consists of two circular rotations, namely, along a circle of radius $\frac{1}{2}(A + B)\chi$ with a constant angular velocity γ and along a circle of radius $\frac{1}{2}(A - B)\bar{\chi}$ with the same angular velocity but in the opposite direction. As a result, each particle describes an ellipse with semi-axes $A|\chi|$ and $B|\chi|$. One such elliptical trajectory, e.g., corresponding to the value $|\chi| = 1$, can be chosen as a solid boundary. Then, the exact solution (3.17) will determine the rotational flow inside the ellipse with semi-axes A and B. In the plane of Lagrangian variables, this motion corresponds to the region $|\chi| \leq 1$.

The flow (3.17) is vortical. Its vorticity, according to Eq. (3.12), is equal to

$$\Omega = \frac{\gamma(A^2 + B^2)}{AB}.$$

The vorticity is constant in the flow region.

Let us complicate the problem and assume that the boundary of the ellipse rotates with a constant angular velocity δ. The motion of fluid particles inside it is determined by the expression

$$W = \frac{1}{2}(A + B)\chi \exp\left[i(\delta + \gamma)t\right] + \frac{1}{2}(A - B)\bar{\chi}\exp[i(\delta - \gamma)t]. \quad (3.18)$$

It is also an exact solution and satisfies the condition of impermeability on a solid surface. The vorticity of motion (3.18) is equal to

$$\Omega = \frac{\gamma(A^2 + B^2)}{AB} + 2\delta. \quad (3.19)$$

This flow is non-stationary. The motion of fluid particles still represents a superposition of two circular rotations, but the rotation frequencies are no longer the same. Fluid trajectories are epicycloids if the frequencies $\delta + \gamma$ and $\delta - \gamma$ have the same sign and hypocycloids if they have different signs. It is noteworthy that if the values of these frequencies are incommensurable, then the trajectories of the particles are not closed.

3.3.4. *Flows with helical particle trajectories*

Let us point out one interesting property of the Lagrange equations. Suppose we know one of their two-dimensional solutions, $X = X_*(a, b, t); Y = Y_*(a, b, t)$. Then, the expressions

$$X = X_*(a, b, t), \quad Y = Y_*(a, b, t), \quad \text{and} \quad Z = c + U(a, b, t), \quad (3.20)$$

where $U(a, b)$ is an arbitrary function, will also be the solution to the equations of fluid dynamics. To prove this, it is enough to substitute these expressions into continuity Eq. (1.11) and note that the spatial momentum Eq. (1.18) in the case

$$Z_{tt} = 0; \quad X_c = 0; \quad Y_c = 0$$

Vortex Motion

is equivalent to a system of equations of two-dimensional fluid dynamics.

Let us choose the flow (3.16) as a known two-dimensional solution. Then, the relations of the form

$$X = a\cos\omega_* t - b\sin\omega_* t; \quad Y = a\sin\omega_* t + b\cos\omega_* t;$$
$$Z = c + U(a,b)t,; \quad \omega_* = \omega_*(|\chi|^2)$$

determine the flow with spiral trajectories of fluid particles. Each particle, in addition to uniform rotation in the XY plane, also participates in inhomogeneous drift along the Z-axis. Such helical motion qualitatively resembles the trajectories along which the medium particles move inside the tornado-like structures.

Particle trajectories will have a similar character if solutions (3.17) or (3.18) are substituted into Eq. (3.20). In the first case, particle trajectories represent "winding" on the surface of an elliptical cylinder, and in the second case, they represent epicycloids (hypocycloids) uniformly elongated in the transverse direction.

3.4. The Kirchhoff Vortex

A Kirchhoff vortex is an elliptical, uniformly vortical region that uniformly rotates in the surrounding potential flow [4,5]. To describe such a vortex solution, it is necessary to overcome two problems, namely, find the type of flow inside the vortex core and connect it with external potential motion. At the flow boundary, the condition of velocity continuity must be fulfilled. The pressure at the boundary will be continuous (see [5], §146).

In his analysis, Kirchhoff started from Eulerian coordinates, and only at the end of the study did he obtain explicit expressions for the coordinates of the trajectories of fluid particles. In contrast to his technique, we initially will start with the Lagrangian description. Let the vortex interior be described by the expression

$$W = \frac{1}{2}(A+B)\chi\exp(i\lambda t) + \frac{1}{2}(A-B)\bar{\chi}; \quad |\chi| \le 1, \qquad (3.21)$$

where λ is a constant frequency. This expression is a special case of Eq. (3.18), where $\delta = \gamma$ and $\lambda = \delta + \gamma$, and obviously satisfies

Eqs. (3.11) and (3.12). The vortex boundary is an ellipse with semi-axes A and B, which rotates with a frequency of $\lambda/2$ around its center. The particles themselves rotate in a circle with twice the frequency around the center, the position of which is determined by the second term (3.21). The Kirchhoff vortex vorticity is equal to

$$\Omega = \frac{\lambda(A+B)^2}{2AB}.$$

This formula can be obtained by direct calculation from both Eqs. (3.12) and (3.19).

To record the potential motion outside the vortex, it is possible to use John's method (see Section 2.7). We assume that in the region $|\chi|^2 > 1$, the variable χ is a parameter. Then, the motion in the region external to the vortex is specified parametrically:

$$W = \frac{1}{2}(A+B)\chi \exp(i\lambda t) + \frac{1}{2}(A-B)\frac{1}{\chi};$$
$$V = V_x + iV_y = \frac{i\lambda(A+B)}{2\bar{\chi}} \exp(i\lambda t), \quad |\chi| > 1. \tag{3.22}$$

These relations determine the potential velocity field $\bar{V}(W,t)$. The function W continuously extends into the unit circle exterior as an analytical function of χ. The function has no branching points since $W_\chi \neq 0$ for $|\chi| > 1$. At the boundary of the sewing vortex and potential flows, the velocity is continuous, $V|_{|\chi|=1} = W_t|_{|\chi|=1}$.

Excluding the parameter, we write the expression for the velocity outside the vortex in explicit (Eulerian) form:

$$\bar{V} = -\frac{i\lambda(A+B)^2}{2(W + \sqrt{W^2 - (A^2 - B^2)\exp(i\lambda t)})}.$$

At "infinity," the velocity decreases as a point vortex field (see Section 2.3):

$$\bar{V}_\infty = -\frac{i\Gamma_*}{2\pi W},$$

where $\Gamma_* = \Omega \cdot S_v$ is the vortex intensity (S_v is the ellipse area).

3.5. Unsteady Motion of a Flat Layer with a Free Boundary

Let us write out Eqs. (1.11) and (1.23) for a plane flow:

$$X_a Y_b - X_b Y_a = D_0(a, b);$$
$$X_{ta} X_b - X_{tb} X_a + Y_{ta} Y_b - Y_{tb} Y_a = D_0 \Omega(a, b). \qquad (3.23)$$

Assume that the function Y does not depend on a. Then, system (3.23) will be reduced and take the following form:

$$\frac{\partial}{\partial t} X_a Y_b = 0; \quad \frac{\partial}{\partial t}(X_{ta} X_b - X_{tb} X_a) = 0.$$

The authors of another study [6] showed that the expressions

$$X = [1 + f(b)t]a + \varphi(b)t \quad \text{and} \quad Y = \int_0^b [1 + f(\mu)t]^{-1} d\mu \qquad (3.24)$$

where f and φ are arbitrary velocities with dimensions of time and velocity, respectively, are the exact solution of these equations.

Let the band $0 \leq b \leq h$ correspond to solution (3.24) in the plane of Lagrangian variables. Then, the solution can be interpreted as the unsteady motion of a flat layer of thickness

$$l(t) = \int_0^h [1 + f(\mu)t]^{-1} d\mu.$$

At the lower boundary of the layer, corresponding to $b = 0$, the impermeability condition ($Y_t = 0$) is fulfilled. With the increase in t and in the case $f(b) > 0$, the layer becomes thinner, so that $l(t) \to 0$ at $t \to \infty$. For $f(b) < 0$, there is such a moment of time t_* that $l(t_*) = 0$.

The pressure inside the fluid, according to Eq. (1.18), is determined by the formula

$$p = -\rho \int_0^b (Y_{tt} Y_b + g Y_b) db,$$

where g is the free-fall acceleration, and is equal to

$$p = -\rho g Y + 2\rho \int_0^b [1 + f(\tau)t]^{-1} \int_0^\tau f^2(\mu)[1 + f(\mu)t]^{-3} d\mu \, d\tau + C(t).$$

At the upper boundary corresponding to $b = h$, it is possible to satisfy the constant pressure condition by choosing a certain value of $C(t)$. Thus, solution (3.24) describes spreading of a flat layer along a solid wall with a free surface. The flow vorticity is equal to

$$\Omega = -af_b - \varphi_b.$$

A layer with constant vorticity Ω_0 is obtained at $f = k = \text{const}$ and is described by the formulas

$$X = (1 + kt)a - \Omega_0 bt, \quad Y = \frac{b}{1 + kt}, \quad \text{and} \quad \Omega_0 = -\varphi_b = \text{const},$$

$$(3.25)$$

and the pressure in the layer has the form

$$p = -\rho g[h - y(1 + kt)] + \frac{\rho k^2}{(1 + kt)^4}[h^2 - (1 + kt)^2 y^2],$$

where y is the vertical Eulerian coordinate. The line $y = h(1 + kt)^{-1}$ is a free boundary.

Due to the linear dependence of X and Y on Lagrangian coordinates, solution (3.25) can be explicitly written in Eulerian coordinates

$$V_x = \frac{kx}{1 + kt} - \Omega_0 y; \quad V_y = -\frac{ky}{1 + kt}.$$

3.6. Spatial Flows with a Velocity Field Linearly Dependent on Coordinates

The equations of fluid dynamics are significantly simplified if we assume that the velocity field is linear in spatial coordinates. The expression for the vector $\vec{R}$, which determines the motion of fluid particles labeled by the Lagrangian vector $\vec{a}$, can be represented as follows:

$$\vec{R} = \hat{A}(t)\vec{a}. \tag{3.26}$$

Here, $\vec{R} = \{X, Y, Z\}$. The elements of the Jacobi matrix $A_{ij} = \partial R_i/\partial a_j$ $(i, j = 1, 2, 3)$ are functions of time only. To determine them, it is necessary to solve continuity Eq. (1.11), which, with allowance for the fact that the Lagrangian variables coincide with the initial coordinates of the particles $(D_0 = 1)$, has the form

$$\det \hat{A} = 1, \tag{3.27}$$

and three momentum Eq. (1.23), written in view of Eqs. (1.26) and (3.26) in the following form:

$$\hat{A}_t^{\mathrm{T}} \hat{A} - \hat{A}^{\mathrm{T}} \hat{A}_t = \hat{S}. \tag{3.28}$$

The elements of the time-independent antisymmetric matrix $\hat{S}$ are the Cauchy invariants. For the flows given by Eq. (3.26), these are some constants. According to the well-known representation of the velocity field, the pressure distribution can be found from Eq. (1.18).

Thus, in general, there are only four conditions for finding the nine components of the matrix $\hat{A}$. However, we will be interested only in some particular solutions of Eqs. (3.27) and (3.28), which can be given a physical interpretation.

3.6.1. *Dirichlet problem*

Historically, Dirichlet [7] was the first to study such flows in order to analyze the properties of self-gravitating rotating ellipsoids. He analyzed under what conditions a flow bounded by the surface of an ellipsoid at any time and described by a velocity field linear in spatial coordinates is possible. Let us outline the essence of his approach for the case of an ordinary (non-gravity) fluid. We will use the flow description method used in a study [8] (and given in another study [6]).

Assume that the pressure inside the fluid ellipsoid is distributed according to the following law:

$$p = \frac{q(t)}{2}(l - \vec{a} \cdot \hat{N}\vec{a}), \tag{3.29}$$

where $\hat{N} = \mathrm{diag}(n_1, n_2, n_3)$ is the diagonal matrix, $n_1 n_2$, and n_3 are the arbitrary numbers, and $\vec{a} = \{a, b, c\}$. The free surface on which the pressure is zero corresponds to a second-order curve

$$l - \vec{a} \cdot \hat{N}\vec{a} = l - (n_1 a^2 + n_2 b^2 + n_3 c^2) = 0.$$

For an ellipsoid, all n_i are positive. Lagrangian coordinates coincide with the initial position of fluid particles and $\hat{A}(0) = \hat{E}, \hat{E}$ is a unit matrix. It follows that at the initial time, the fluid volume has the shape of an ellipsoid. In Eulerian variables, the shape of the free boundary is determined by the equation

$$\hat{A}^{-1}\vec{R}\hat{N}\hat{A}^{-1}\vec{R} = l.$$

It is obvious that the surface specified by this quadratic shape will remain an ellipsoid at all subsequent times.

With allowance for Eqs. (3.26) and (3.29), Eq. (1.17) can be written as

$$\hat{A}_{tt} = q(\hat{A}^{\mathrm{T}})^{-1}\hat{N}, \tag{3.30}$$

which should be supplemented with an initial condition

$$A_t(0) = A_0, \quad Sp\, A_0 = 0.$$

Here, A_0 is an arbitrary constant matrix. The last condition follows from the condition that the divergence of the Eulerian velocity field is equal to zero at the initial time.

The Dirichlet problem is commonly solved in Eulerian variables. Lagrangian formalism "does not bring much benefit in the study of specific cases of fluctuations" [9]. At the same time, a number of results on the dynamics of non-stationary, non-gravity ellipsoids can be obtained precisely in Lagrangian variables.

3.6.2. *Ovsyannikov ellipsoid*

Let us choose the matrices $\hat{A}$ and $\hat{N}$ in the following form [8]:

$$\hat{A} = \begin{pmatrix} \alpha & -n & 0 \\ n & \alpha & 0 \\ 0 & 0 & m \end{pmatrix}, \quad \hat{N} = \hat{E} = \begin{pmatrix} 1 & 0 & 0 \\ 0 & 1 & 0 \\ 0 & 0 & 1 \end{pmatrix}. \tag{3.31}$$

In the expression for pressure (3.29), we assume $l = 1$. The coordinates of the initial position of the fluid particles coincide with the Lagrangian ones,

$$\alpha(0) = m(0) = 1, \quad n(0) = 0.$$

Substituting the expression for the matrix $\hat{A}$ into continuity Eq. (3.27), we find that the elements of the matrix are connected by the relation

$$(\alpha^2 + n^2)m = 1.$$

With allowance for this equality, the matrix inverse of the transposed matrix $\hat{A}^{\mathrm{T}}$ is written as

$$(\hat{A}^{\mathrm{T}})^{-1} = \begin{pmatrix} \alpha m & -nm & 0 \\ nm & \alpha m & 0 \\ 0 & 0 & m^{-1} \end{pmatrix},$$

and the system of Eq. (3.30) takes the form

$$\alpha_{tt} = q\alpha m, \tag{3.32}$$

$$n_{tt} = qnm, \tag{3.33}$$

$$m_{tt} = qm^{-1}. \tag{3.34}$$

To solve the system of Eqs. (3.32)–(3.34), we assume

$$\alpha = m^{-1/2} \cos\left(\omega \int_0^t m(\tau)d\tau\right); \quad n = m^{-1/2} \sin\left(\omega \int_0^t m(\tau)d\tau\right),$$

where ω is the constant. Equations (3.32) and (3.33) will be satisfied if we set

$$q = \frac{3m_t^2 - 4\omega^2 m^4}{2(1 + 2m^3)}.$$

For this reason, Eq. (3.34) will take the form

$$m_{tt} = \frac{3m_t^2 - 4\omega^2 m^4}{2m(1 + 2m^3)}.$$

Lowering its order, we find that the function $m(t)$ satisfies the equation

$$m_t^2 = \frac{12b^2 m^3[1 + \lambda(1 - m)]}{1 + 2m^3}, \quad \lambda = \frac{\omega^2}{3b^2}, \tag{3.35}$$

where the parameter $b = -\alpha_t(0) = m_t(0)/2$, so that the matrix A_0, which determines the initial flow velocity, has the form

$$A_0 = \begin{pmatrix} -b & -\omega & 0 \\ \omega & b & 0 \\ 0 & 0 & 2b \end{pmatrix}.$$

66   *Analytical Fluid Dynamics in Lagrangian Variables*

According to this matrix, the initial flow vorticity is equal to 2ω, and the vorticity vector is directed along the vertical axis z. Since the Cauchy invariants are equal to the magnitude of the initial vorticity (see Section 1.4), the relations

$$S_1 = S_2 = 0 \quad \text{and} \quad S_3 = 2\omega$$

are valid for them. The coordinates of the fluid particles inside the ellipsoid change according to the following law:

$$X = \alpha a - nb, \quad Y = na + \alpha b, \quad Z = mc.$$

The vorticity vector in accordance with Eq. (1.38) is determined by the expression

$$\vec{\Omega} = 2\omega \vec{R}_c = 2\omega Z_c \vec{z}^0 = 2\omega m(t)\vec{z}^0.$$

It is directed vertically all the time and its magnitude is time-dependent.

We will consider the vortex flow mode ($\omega \neq 0$) and assume that $b > 0$. Then, $m_t(0) > 0$, and the value of $m(t)$ increases with increasing t, but at a critical value of $m_* = 1 + 1/\lambda$, the quantity m_t vanishes. This happens at the time t_* determined by the formula

$$2b\sqrt{3}t_* = \int_1^{m_*} \tau^{-3/2} \sqrt{\frac{1 + 2\tau^3}{1 + \lambda(1 - \tau)}} d\tau.$$

When $t > t_*$, the following inequalities must be fulfilled:

$$m < m_* \quad \text{and} \quad m_t < 0$$

In this case, the $m(t)$ dependence is determined by the equations

$$2b\sqrt{3}t = \int_1^{m_*} \tau^{-3/2} \sqrt{\frac{1 + 2\tau^3}{1 + \lambda(1 - \tau)}} d\tau, \quad 0 \leq t \leq t_*,$$

$$2b\sqrt{3}(t - t_*) = \int_m^{m_*} \tau^{-3/2} \sqrt{\frac{1 + 2\tau^3}{1 + \lambda(1 - \tau)}} d\tau, \quad t \geq t_*. \tag{3.36}$$

It can be seen from the second Eq. (3.36) that $m \to 0$ at $t \to \infty$. If $b < 0$, then the function m monotonically decreases at $t > 0$.

At the initial time, the fluid volume is a ball; at later moments, it deforms, remaining an ellipsoid. The equation of its boundary can be written as follows:

$$m(x^2 + y^2) + m^{-2}z^2 = 1.$$

If $b > 0$, then at $0 \leq t \leq t_*$, the ball is pulled into an ellipsoid of rotation with the z-axis until its major (directed along z) semi-axis takes the value $m = m_*$ at the time $t = t_*$. After that, the ellipsoid starts to flatten vertically, and at the time $t = 2t_*$ it passes the stage of the initial sphere and then continues to shrink to the plane $z = 0$.

Now, let $\omega = 0$. In this case, S_3 and Ω are equal to zero and, therefore, the flow is potential. The form of the matrix $\hat{A}$ is also reduced: Since now $n = 0$, it becomes diagonal, and the relation $m = \alpha^{-2}$ is fulfilled between its non-zero elements. The function $\alpha(t)$ is determined by the expression

$$\int_1^\alpha \alpha^{-3}(2 + \alpha^6)^{1/2} d\alpha = kt \ (k = b\sqrt{3}). \tag{3.37}$$

With increasing time, the unit ball deforms into an ellipsoid of rotation, with the semi-axes equal to $\alpha, \alpha, \alpha^{-2}$. If $k < 0$, then at $t \to \infty$, according to Eq. (3.37), $\alpha \to 0$. The ellipsoid extends along the z-axis, becoming thinner. At large t, it has the form of a needle. The velocity of its end is equal to

$$m_t(t) = 2|k|(2 + \alpha^6)^{-1/2}.$$

It increases with time, i.e., if at $t = 0$ its value is $2|k|/\sqrt{3}$, then at $t \to \infty$, when $\alpha = 0$, it reaches the value $\sqrt{2}|k|$. If $k > 0$, then the ellipsoid given by the equation

$$\alpha^{-2}(x^2 + y^2) + \alpha^4 z^2 = 1$$

at $t \to \infty$ is flattened to the plane $z = 0$.

The above-mentioned solution was obtained by L.V. Ovsyannikov [8] and represents an exceptional case of the analytical solution to the non-stationary Dirichlet problem. Developing this approach, O.M. Lavrentieva obtained some fairly general results on the dynamics of non-stationary ellipsoids with a free boundary [10]. In particular, she pointed out the inequality that the diameter of the ellipsoid should satisfy for some specific representations of the matrices $\hat{A}_0$ and N.

References

[1] Batchelor, G.K. (1967). *An Introduction to Fluid Dynamics* (Cambridge University Press, Cambridge).

[2] Abrashkin, A.A. and Yakubovich, E.I. (1984). Two-dimensional vortex flows of an ideal fluid. *Soviet Phys. Dokl.*, 29, pp. 370–371.

[3] Abrashkin, A.A. and Yakubovich, E.I. (1985). Nonstationary vortex flows of an ideal incompressible fluid. *J. Appl. Mech. Tech. Phys.*, 26, pp. 202–208.

[4] Lamb, H. (1932). *Hydrodynamics*, 6th edn. (Cambridge University Press, Cambridge).

[5] Kirchhoff, G. (1883). *Mechanik* (Teubner, Leipzig).

[6] Andreev, V.K. and Rodionov, A.A. (1988). *Diff. Equations*, 24, pp. 1577–1586.

[7] Chandrasekhar, S. (1969). *Ellipsoidal Figures of Equilibrium* (Yale University Press, New Haven and London).

[8] Ovsyannikov, L.V. (1967). *General Equations and Examples. The Problem of Unsteady Fluid Motion with a Free Boundary* [in Russian], (Nauka, Novosibirsk).

[9] Kondrat'ev, B.P. (2003). *Theory of Potential and Figures of Equilibrium* [in Russian], (Inst. Komp. Issled., Moscow, Izhevsk).

[10] Lavrent'eva, O.M. (1980). Motion of a fluid ellipsoid. *Dokl. Akad. Nauk SSSR*, 253(4), pp. 828–831 [in Russian].

Part II
Waves on Water

Chapter 4

Potential Approximation

"Stokes' studies of waves on water (first published [1] in 1847)
marked the beginning of the nonlinear theory of
dispersing waves."

— Whitham G.B., *Linear and Nonlinear Waves* [2]

The classical problem of describing gravity waves on the surface of a
fluid comes down to solving the Laplace equation for the flow potential. However, in contrast to the standard problems of mathematical
physics, the boundary conditions are set on a free surface, the form of
which is unknown in advance. This circumstance significantly complicates the motion analysis. As a result, no exact solution to this
problem has been obtained till the present time.

Stokes solved this nonlinear wave problem for a stationary periodic wave [1] by expanding the corresponding values of the potential flow into a Taylor series around the average (or stable) surface
height. As a result, the boundary conditions were transformed into
conditions at an average (stable) surface height, which is fixed and
known. As a next step, all the desired functions were represented
using a series of perturbations with respect to a small parameter of
wave steepness. This procedure was called the Stokes expansion and
the constructed approximate nonlinear solution was called the Stokes
wave.

The representation of the Stokes wave in Lagrangian variables
is shown in the following. In the Lagrangian description, the free
boundary can be given by the ratio $b = 0$. One of the fundamental

72   *Analytical Fluid Dynamics in Lagrangian Variables*

difficulties of the Eulerian description of waves on water is removed, i.e., the solution to the problem is now sought at a known boundary, which greatly simplifies the construction of Stokes expansions.

4.1. Stokes Wave

Consider a stationary periodic wave propagating along the free surface of a fluid at a constant speed in the positive direction of the OX axis. The equations of two-dimensional fluid dynamics can be written in the form

$$X_{tt} = -P_a Y_b + P_b Y_a, \quad Y_{tt} = -P_b X_a + P_a X_b,$$

$$X_a Y_b - X_b Y_a = 1, \quad \text{and} \quad P = \frac{p}{\rho} + gY. \tag{4.1}$$

This form of writing is obtained from the systems of Eqs. (1.11) and (1.18) if we assume

$$Z = Z_0 = \text{const}, \quad D_0 = 1, \quad \text{and} \quad H = gy$$

and solve them relative to X_{tt} and Y_{tt}. When studying the wave motion of a fluid, it is convenient to switch to a reference frame traveling at the wave propagation velocity c, where the flow is stationary.

4.1.1. *The method of modified Lagrangian coordinates*

In Lagrangian variables, the stationary flow has the form

$$X = X(q, b), \quad Y = Y(q, b), \quad q = a + \sigma(b)t, \tag{4.2}$$

where $\sigma(b)$ is an arbitrary function. The easiest way to verify the validity of this statement is to write down the flow velocity field (4.2) in Eulerian coordinates. Since the Lagrangian velocity components $X_t = \sigma X_q$ and $Y_t = \sigma Y_q$, like the functions X and Y, depend only on two variables, q and b, the Eulerian velocity field $X_t(X, Y), Y_t(X, Y)$ does not explicitly depend on time and, therefore, describes a stationary flow. Note that the coordinate q is no longer the mark of an individual particle. Due to this, the proposed approach can no longer be called Lagrangian. The coordinates q and b were introduced for the

Potential Approximation

first time in another study [3] and referred to as modified Lagrangian variables.

Using the variables q and b, Eq. (4.1) will be written in the form

$$X_q Y_b - X_b Y_q = 1,$$
$$\sigma^2 X_{qq} = -P_q Y_b + P_b Y_q; \quad \sigma^2 Y_{qq} = -P_b X_q + P_q X_b. \tag{4.3}$$

Assume that

$$X = q + \xi, \quad Y = b + \eta,$$

where the functions ξ and η mean periodic wave disturbances of the fluid particle trajectory from modified Lagrangian coordinates. Equation (4.3) will be rewritten as follows:

$$\xi_q + \eta_b = -\frac{D(\xi, \eta)}{D(q, b)}, \tag{4.4}$$

$$\sigma^2 \xi_{qq} = -P_q + \frac{D(\eta, P)}{D(q, b)}, \tag{4.5}$$

$$\sigma^2 \eta_{qq} = -P_b - \frac{D(\xi, P)}{D(q, b)}. \tag{4.6}$$

These equations should be supplemented with boundary conditions. In the depth, the vertical velocity should drop to zero,

$$Y_t = \sigma Y_q = \sigma \eta_q \to 0 \quad \text{for } b \to -\infty, \tag{4.7}$$

and on the free surface $(b = 0)$, the pressure should be constant, i.e.,

$$P(q, 0) - g\eta(q, 0) = \frac{p_0}{\rho} = \text{const.} \tag{4.8}$$

Conditions (4.7) and (4.8) should also be supplemented with the requirement that the OX axis correspond to the average fluid level. This requirement can be written in the form

$$\int_0^\lambda Y\, dX \bigg|_{b=0} = \int_0^\lambda \eta(1 + \xi_q) dq \bigg|_{b=0} = 0, \tag{4.9}$$

where λ is the wavelength. Since the fluid depth is infinite, it should be natural to assume that there are no wave disturbances at the bottom and the horizontal velocity related to them also vanishes.

74   *Analytical Fluid Dynamics in Lagrangian Variables*

In this case, the motion near the bottom is described by the expressions

$$X = q \quad \text{and} \quad Y = b.$$

They describe a shear flow with $\sigma(b)$ profile. The magnitude of this function at the bottom is equal to the stationary wave velocity with a minus sign,

$$\sigma(-\infty) = -c. \tag{4.10}$$

In another way, this implies that in a laboratory reference frame, the velocity at the bottom is zero. The difference between the shear flow $\sigma(b)$ and the velocity at the bottom means the particle drift velocity,

$$u(b) = \sigma(b) - \sigma(-\infty) = \sigma(b) + c. \tag{4.11}$$

Modified variables are convenient to use in cases where fluid particles drift in a certain direction. The point of introducing new coordinates is that in the representation for the current coordinates of a fluid particle, a drift, which is non-uniform in the vertical coordinate b, is eliminated (more precisely, it is "driven" inside the variable q). Thus, in the representation of modified Lagrangian variables, each fluid particle participates only in oscillatory motion relative to a certain center, and if the deviation of the particle from the center is small compared to the wavelength, the problem of such a wave admits an effective solution using the perturbation method. An attempt to construct in this way a solution for a weakly nonlinear wave in ordinary Lagrangian coordinates in the presence of a drift stream leads to the appearance of secular terms. The new coordinate q is no longer related to a specific fluid particle; it is some auxiliary variable. It is introduced in such a way that perturbations of a given shear flow depend on it periodically. This coordinate plays the same role in problems about waves on water as the horizontal coordinate in the standard Eulerian description. Taking this into consideration, the method of modified coordinates cannot be called purely Lagrangian; it is of a mixed nature.

Our aim is to provide a description of weakly nonlinear vortex-free waves. The expression for vorticity (see Eq. (1.32)) has the form

$$\Omega = -\sigma'(1 + \xi_q) + \sigma(\eta_{qq} - \xi_{qb}) + \frac{D(\sigma\xi_q, \xi)}{D(q, b)} + \frac{D(\sigma\eta_q, \eta)}{D(q, b)}.$$

Obviously, this expression should be zero. Due to the stationarity of the motion under consideration, the amount of vorticity depends only on the coordinate b. Therefore, this relation can be simplified by eliminating the q-periodic terms through averaging. The wave potentiality condition will then be written as follows:

$$\Omega = -\sigma' + \overline{\frac{D(\sigma\xi_q, \xi)}{D(q, b)}} + \overline{\frac{D(\sigma\eta_q, \eta)}{D(q, b)}} = 0. \tag{4.12}$$

Here, the bar is the averaging over the variable q at the wavelength $\lambda = 2\pi/k$ and k is the wave number.

Let us represent the unknown functions ξ, η, u, H, and c as series in powers of the small parameter ε:

$$\{\xi, \eta, u\} = \sum_{n=1}^{\infty} \varepsilon^n \{\xi_n, \eta_n, u_n\};$$

$$\{P, c\} = \left\{\frac{p_0}{\rho}, c_0\right\} + \sum_{n=0}^{\infty} \varepsilon^n \{P_n, c_n\}. \tag{4.13}$$

Equation (4.11) can be rewritten as follows:

$$\sigma = -c_0 + \sum_{n=1}^{\infty} \varepsilon^n (u_n - c_n) = \sigma_0 + \sum_{n=1}^{\infty} \varepsilon^n \sigma_n. \tag{4.14}$$

The physical meaning of the parameter ε will be indicated in the following.

4.1.2. *Linear wave*

The solution of the linear approximation has the following form:

$$\xi_1 = -\frac{1}{k} e^{kb} \sin kq; \quad \eta_1 = \frac{1}{k} e^{kb} \cos kq; \quad H_1 = \sigma_0^2 e^{kb} \cos kq.$$

Since in the flow region $b \le 0$ the oscillation velocities $\sigma_0 \xi_{1q}$ and $\sigma_0 \eta_{1q}$ of fluid particles decrease exponentially with the depth, the condition (4.7) of the absence of vertical velocity at the bottom is fulfilled. For constant pressure, the equality

$$P_1(q, 0) = g\eta_1(q, 0)$$

should be satisfied (see Eq. (4.8)). It follows that $\sigma_0^2 = g/k = c_0^2$. This is the square of the phase velocity of a linear potential wave, whose frequency is equal to $\sqrt{gk}$.

From the potentiality condition (4.12), we find that $\sigma'_1 = 0$ or, taking into account Eq. (4.10), $\sigma_1 = -c_1$. The quantity c_1 cannot be determined in the linear approximation. It can be found in the quadratic approximation. Therefore, the expression for the linear Stokes wave in Lagrangian variables can be written in the form

$$X = a - c_0 t - \varepsilon c_1 t - \varepsilon \cdot \frac{1}{k} e^{kb} \sin k(a - c_0 t);$$

$$Y = b + \varepsilon \cdot \frac{1}{k} e^{kb} \cos k(a - c_0 t).$$

It can be seen from these relations that the small parameter should be chosen in the form $\varepsilon = kA$, where A is the wave amplitude.

We emphasize that the zero vorticity condition is not sufficient to fully determine the function σ_1. The latter is found up to a constant, which should be calculated in the next approximation. This feature of calculating σ_n will also be preserved in higher orders of perturbation theory.

In a laboratory reference frame, the solution of a linear problem has the form

$$X = a - \varepsilon \cdot \frac{1}{k} e^{kb} \sin k(a - c_0 t);$$

$$Y = b + \varepsilon \cdot \frac{1}{k} e^{kb} \cos k(a - c_0 t).$$

$$(4.15)$$

The wave propagates to the right with velocity c_0. In this case, the fluid particles rotate in a circle,

$$(X - a)^2 + (Y - b)^2 = \frac{\varepsilon^2}{k^2} \exp 2kb.$$

There is no drift of fluid particles in the linear approximation: $u_1 = 0$ (see Eq. (4.11)).

In the Lagrangian description, the trajectories of fluid particles are explicitly determined. In order to determine the coordinates X and Y of the trajectories of particles using the Eulerian approach, it is necessary to solve two first-order differential equations. There should be $\frac{dX}{dt}$ and $\frac{dY}{dt}$, respectively, on the left-hand sides of this system and the expressions for horizontal and vertical velocity on the right-hand sides. For a linear wave in deep water, this system is nonlinear and non-integrable [4]. However, it can be integrated if the values of X and Y on the right-hand sides are replaced by their average values (coinciding with the initial position of the particles). This gives the classical result: Fluid particles in a linear potential wave in deep water move in a circle. However, quite recently, authors of other studies [5–7] drew attention to an interesting property of solutions of a complete nonlinear system for X and Y: None of these solutions describes the motion with closed particle trajectories. The drift of fluid particles (the quadratic approximation effect) in the Eulerian representation seems to be present (potentially) even in a linear wave.

4.1.3. Stokes drift: Nonlinear dispersion relation

In the quadratic approximation in a laboratory reference frame, the solution is written as follows:

$$X = a - \varepsilon \cdot \frac{1}{k}e^{kb}\sin k(a - c_0 t) + \varepsilon^2 c_0 e^{2kb}t + O(\varepsilon^3); \qquad (4.16)$$

$$Y = b + \varepsilon \cdot \frac{1}{k}e^{kb}\cos k(a - c_0 t) + \varepsilon^2 \cdot \frac{1}{2k}e^{2kb} + O(\varepsilon^3);$$

$$c_1 = 0. \qquad (4.17)$$

Specific details of the calculations can be followed by addressing Section 6.1.1, where this problem is considered with allowance for the vorticity of wave disturbances. Equations (4.16) and (4.17), as well as all subsequent expressions in this section, are obtained from the formulas in Section 6.1.1 by zeroing the vortices of the corresponding approximation.

Equations (4.16) and (4.17) were obtained by Stokes [1]; however, the term $\varepsilon^2 e^{2kb}/2k$ was absent in the second equation in his original work. Taking into account this term ensures that the average fluid level remains at zero (condition (4.9)). The third term on the right-hand side of Eq. (4.16) describes the non-uniform vertical drift of fluid particles (Stokes drift): $U_s = \varepsilon^2 c_0 e^{2kb}$.

We will not derive solutions of the cubic approximation, but we give the form of a nonlinear dispersion relation obtained for the first time by Stokes [1]:

$$c = c_0 \left[1 + \frac{1}{2}(kA)^2 \right]; \quad c_0 = \sqrt{\frac{g}{k}}. \tag{4.18}$$

The phase velocity of the wave depends on the wave number (frequency) and amplitude; that is to say, the Stokes wave, on the one hand, has dispersion and, on the other hand, is a nonlinear oscillation. This explains the fact that the Stokes wave theory marked the beginning of the development of the nonlinear theory of dispersing waves (see the epigraph to the chapter).

4.1.4. *Overview of other works*

(a) If we use ordinary Lagrangian variables, then, starting from the second approximation, secular terms appear in the solution [1,8] (see Eq. (4.16)). They are due to the presence of non-uniform drift of fluid particles (Stokes drift). The introduction of modified Lagrangian coordinates makes it possible to construct a time-uniform solution for all approximations, including higher-order ones. The authors of a study [9] independently came to the same form of description. For waves in a fluid of finite depth, they found the Stokes expansion up to the fifth order inclusive.

(b) We have written down the stationary flow using modified coordinates. However, other forms of representing stationary flows are possible in Lagrangian coordinates. Clamond [10] used a coordinate of the form $q \cdot K(b)$ in combination with the coordinate b. As in our method, the stream function in this case depends only on the b coordinate. In another study [10], the coordinates providing this property of flows were called simplified. They are convenient to employ for solving wave problems. Thus, in that analysis [10], Stokes expansions up to the seventh order for waves in deep water and up to cubic terms for waves in a fluid of finite depth are found. We will use modified Lagrangian variables in this book.

However, in general, the stream function of a steady-state flow may depend on two Lagrangian coordinates. An example

Potential Approximation

of such motion is a two-dimensional flow near the critical point (Section 2.2.1), for which $\psi = ab$.

(c) A method for reducing the continuity equation for two-dimensional flows to a linear equation similar to the zero divergence condition in Eulerian variables is proposed [11]. Let us represent the coordinates of a fluid particle in the following form:

$$X(a,b) = a + \varepsilon\xi(\alpha,\gamma), \quad Y(a,b) = b + \varepsilon\eta(\alpha,\gamma), \qquad (4.19)$$

where ε is the dimensionless displacement amplitude, and the displacement vector (ξ,η) of the particle from its initial position $(X_0, Y_0) = (a,b)$ is a function of the new variables

$$\alpha = \frac{1}{2}[a + X(a,b)], \quad \gamma = \frac{1}{2}[b + Y(a,b)]. \qquad (4.20)$$

The variables (α,γ) mean medium values between the equilibrium and the current position of the fluid particle. The coordinates of the trajectory of the particles and their corresponding markers as functions of (α,γ) can be written in the form

$$X = \alpha + \frac{\varepsilon}{2}\xi(\alpha,\gamma), \quad Y = \gamma + \frac{\varepsilon}{2}\eta(\alpha,\gamma); \qquad (4.21)$$

$$a = \alpha - \frac{\varepsilon}{2}\xi(\alpha,\gamma), \quad b = \gamma - \frac{\varepsilon}{2}\eta(\alpha,\gamma). \qquad (4.22)$$

Using these representations, the continuity equation can be written in the form

$$D = \frac{D(X,Y)}{D(\alpha,\gamma)} \bigg/ \frac{D(a,b)}{D(\alpha,\gamma)} = 1, \qquad (4.23)$$

where

$$\frac{D(X,Y)}{D(\alpha,\gamma)} = 1 + \varepsilon(\xi_\alpha + \eta_\gamma) + \frac{\varepsilon^2}{4}(\xi_\alpha\eta_\gamma - \xi_\gamma\eta_\alpha);$$

$$\frac{D(a,b)}{D(\alpha,\gamma)} = 1 - \varepsilon(\xi_\alpha + \eta_\gamma) + \frac{\varepsilon^2}{4}(\xi_\alpha\eta_\gamma - \xi_\gamma\eta_\alpha).$$

Substituting these relations into Eq. (4.23), we obtain the linear form of the continuity equation

$$\frac{\partial\xi}{\partial\alpha} + \frac{\partial\eta}{\partial\gamma} = 0. \qquad (4.24)$$

The particle displacements ξ and η can be described by a single function $\Psi(\alpha, \gamma)$, which is similar to the stream function for the Eulerian approach and is defined as

$$\xi(\alpha, \gamma) = -\frac{\partial \Psi}{\partial \gamma}, \quad \eta(\alpha, \gamma) = \frac{\partial \Psi}{\partial \alpha}.$$

Equation (4.24) is an exact result, valid for any ε. But other authors [11], assuming $\varepsilon \ll 1$, constructed an iterative scheme for solving the continuity equation.

The mapping $(a, b) \to (X, Y)$ should be considered as a combination of two mappings, namely, $(a, b) \to (\alpha, \gamma)$ and the following one, $(\alpha, \gamma) \to (X, Y)$. The variables (α, γ) in accordance with Eq. (4.22) are specified by the transform

$$\alpha = a + \frac{\varepsilon}{2}\xi(\alpha, \gamma), \quad \gamma = b + \frac{\varepsilon}{2}\eta(\alpha, \gamma). \qquad (4.25)$$

As the initial approximation, we choose $(\alpha_0, \gamma_0) = (a, b)$. Substituting these values into the right-hand sides of Eq. (4.25), we find (ξ_1, η_1). Expressions for the next approximations $(\xi_m, \eta_m), m > 1$ are sought in a similar way. The corresponding functions (X_m, Y_m) are found through Eq. (4.21).

The aforementioned authors [11] applied this procedure to the analysis of stationary periodic waves in deep water and obtained a Lagrangian asymptotic solution of the fifth order.

4.2. Wave Train of Surface Gravity Waves

Consider a train of gravity waves traveling along the deep water surface. The equations of fluid dynamics describing the non-stationary potential motion of a fluid will be chosen in the form of Eqs. (2.6) and (2.7). We assume that $b = 0$ corresponds to the free surface and $b \to -\infty$ to the bottom. The boundary conditions include the requirements of no leakage at the bottom, where $Y_t \to 0$, and constant pressure on the free surface.

In complex variables (3.2), Eqs. (2.6) and (2.7) will take the form

$$\frac{D(W, \overline{W})}{D(a, b)} = -2i; \qquad (4.26)$$

$$\mathrm{Re}\frac{D(W_t, \overline{W})}{D(a, b)} = 0. \qquad (4.27)$$

Potential Approximation 81

The system of momentum Eq. (4.1) will be written in the form of a single equation

$$W_{tt} = -ig + i\rho^{-1}\frac{D(p, W)}{D(a, b)}. \tag{4.28}$$

Equations (4.26) and (4.27) are used to find the trajectory complex coordinate and Eq. (4.28) to determine the pressure.

Let us employ the multiscale method [12]. We write the function W as follows:

$$W = a_0 + ib + w(a_n, b, t_n), \quad a_n = \varepsilon^n a, \quad t_n = \varepsilon^n t; \quad n = 0, 1, 2, \tag{4.29}$$

where ε is a small parameter of wave steepness, and the functions w and p are represented as series with respect to the parameter

$$w = \sum_{n=1} \varepsilon^n w_n; \quad p = p_0 - \rho g b + \sum_{n=1} \varepsilon^n p_n. \tag{4.30}$$

In the expression for pressure, the term with hydrostatic pressure is separately highlighted and p_0 is the constant atmospheric pressure on the surface of the fluid, which can be taken equal to zero.

Substitute Eqs. (4.29) and (4.30) into Eqs. (4.26) and (4.27). In the first approximation, the solution has the following form:

$$w_1 = A(a_1, a_2, t_1, t_2)\exp[i(ka_0 - \omega t_0) + kb] + \psi_1(a_1, a_2, b, t_1, t_2);$$
$$\omega^2 = gk. \tag{4.31}$$

Hereafter, A is the complex amplitude of the wave traveling to the right. The function ψ_1 is real, and its form should be determined when considering the next approximation. Equation 4.31 describes wave motion in a laboratory reference frame. It consists of the oscillatory motion of fluid particles in a circle and the average flow.

We will not give a detailed description of the calculations but will limit ourselves to indicating the main results in higher-order approximations. From the solutions of the second approximation, a pair of equations follows:

$$A_{t_1} + c_g A_{a_1} = 0, \tag{4.32}$$

$$\psi_{1t_1} = k\omega|A|^2 \exp 2kb. \tag{4.33}$$

Here, $c_g = g/2\omega$ is the group velocity of linear gravity waves. Note that the right-hand side in Eq. (4.33) coincides with the Stokes drift, but now the amplitude A is a function of time and coordinates.

Using Eqs. (4.32) and (4.33), in the third approximation, we come to the following evolution equation:

$$i A_{t_2} - \frac{\omega}{8k^2} A_{a_1 a_1} - \frac{1}{2}\omega k^2 |A|^2 A = 0. \qquad (4.34)$$

This is the nonlinear Schrödinger equation (NLSE). It is written in a reference frame moving with group velocity c_g. Note that the expression for the nonlinear term is found in the relation

$$\left(2k^2 \int_{-\infty}^{0} \psi_{1t_1} e^{2kb}\, db \right) A = \frac{1}{2}\omega k^2 |A|^2 A,$$

i.e., it is determined by the type of drift current of the first approximation. For potential waves in deep water, the NLSE was first derived by Zakharov using the Hamiltonian formalism [13]. Hasimoto and Ono [14] and Davi [15] independently obtained the same result using the multiscale expansion method, and Yuen and Lake obtained it based on the averaged Lagrangian method [16]. Here, we have indicated a way to derive the NLSE in Lagrangian coordinates. This equation is integrable, and many of its exact wave solutions are known (see [17–20]).

In order to write Eq. (4.34) in Eulerian variables, it is necessary to express the horizontal Lagrangian coordinate a in terms of the horizontal Eulerian coordinate X. Since from Eqs. 4.29 and (4.30), it follows that

$$X = a + \varepsilon \mathrm{Re}\left(w_1 + \sum_{n=2} \varepsilon^{n-1} w_n \right) = a + O(\varepsilon),$$

then to switch to the Eulerian form of writing, one should simply replace the Lagrangian coordinate with the corresponding Eulerian variable $(a_n \to X_n)$. Obviously, the inverse substitution of coordinates is also valid; thus, all known solutions of the NLSE in Eulerian variables can also be written in Lagrangian variables in a similar way.

The heuristic derivation of the NLSE based on the nonlinear dispersion relation for the Stokes wave (4.18) is very popular in physics.

Potential Approximation

We rewrite it as

$$\omega = \sqrt{gk}\left(1 + \frac{1}{2}k^2A^2\right).$$
(4.35)

For a narrow wave train ($\Delta k \ll k$), Eq. (4.35) can be expanded in the vicinity of a constant value k_0 (the center of the spectral interval of the train), keeping second-order terms with respect to wave number perturbations and nonlinearity. Then, the perturbations of the wave number k' and frequency ω' satisfy the equation

$$\omega' - \frac{\omega_0}{2k_0}k' + \frac{\omega_0}{8k_0^2}k'^2 - \frac{1}{2}\omega_0 k_0^2 A^2 = 0; \quad \omega_0 = \sqrt{gk}.$$
(4.36)

Considering the frequency and the wave number in Eq. (4.36) as operators in accordance with the relations

$$-i\omega' \to \frac{\partial}{\partial t} \quad \text{and} \quad ik' \to \frac{\partial}{\partial x},$$
(4.37)

we obtain the NLSE [21]:

$$i\left(\frac{\partial A}{\partial t} + c_g\frac{\partial A}{\partial X}\right) - \frac{\omega_0}{8k_0^2}\frac{\partial^2 A}{\partial X^2} - \frac{1}{2}\omega_0 k_0^2 |A|^2 A = 0.$$

This equation is similar to Eq. (4.34) if we write it in a reference frame moving with group velocity to the right and make substitutions

$$A \to \frac{A}{\varepsilon}, \quad t_2 \to \varepsilon^2 t, \quad \text{and} \quad a_1 \to \varepsilon X.$$

This heuristic conclusion clearly indicates that the nonlinearity coefficient in the NLSE coincides with the nonlinear correction to the dispersion relation for linear waves.

Let us rewrite the dispersion relation (4.35) in a slightly different form

$$\omega - k \cdot u = \omega_0.$$
(4.38)

Here, the value $u = \frac{1}{2}\omega_0 k A^2$ corresponding to the nonlinear correction to the phase velocity was introduced. It follows from Eq. (4.38)

84   *Analytical Fluid Dynamics in Lagrangian Variables*

that this velocity can be interpreted as a surface flow providing a Doppler frequency shift. This velocity satisfies the relations

$$u = k \int_{-\infty}^{0} U_s(b)db = k \int_{-\infty}^{0} U_s(y)dy,$$

where $U_s(b)$ and $U_s(y) = c_0 e^{2ky}(kA)^2$ are the expressions for Stokes drift in Lagrangian (see Section 4.1.3) and Eulerian representations, respectively [1, 22]. The velocity u means the flow rate through a transverse area of unit length multiplied by k. With allowance for the fact that the drift velocity decreases exponentially with the depth and is significant only in the near-surface layer with a thickness of the order of the wavelength, it can be concluded that the drift velocity is approximately equal to the vertical-averaged Stokes drift velocity.

4.3.   Dam Failure Problem

Stoker's book [23] presents Pohle's results [24, 25] devoted to the method of studying non-stationary potential flows in the initial period of motion (for a short observation time). Let us rewrite the equations of two-dimensional fluid dynamics:

$$X_a Y_b - X_b Y_a = 1; \tag{4.39}$$

$$X_{tt} X_a + (Y_{tt} + g)Y_a + \frac{1}{\rho} p_a = 0; \tag{4.40}$$

$$X_{tt} X_b + (Y_{tt} + g)Y_b + \frac{1}{\rho} p_b = 0. \tag{4.41}$$

For the potential flow, the consequence of the last two equations is the expression

$$X_{ta} X_b - X_{tb} X_a + Y_{ta} Y_b - Y_{tb} Y_a = 0, \tag{4.42}$$

(see Section (1.4)). Pohle's method assumes that unknown functions X, Y, and p are represented as the following time series expansions

$$X(a, b; t) = a + X^{(1)}(a, b)t + X^{(2)}(a, b)t^2 + \cdots$$

$$Y(a, b; t) = b + Y^{(1)}(a, b)t + Y^{(2)}(a, b)t^2 + \cdots \tag{4.43}$$

$$p(a, b; t) = p^{(0)}(a, b) + p^{(1)}(a, b)t + p^{(2)}(a, b)t^2 + \cdots,$$

Potential Approximation

where the coefficients $X^{(n)}, Y^{(n)}, p^{(n-1)}, n = 1, 2, \ldots$ are functions of Lagrangian coordinates. Equation (4.43) describes flows for which the initial positions of the particles are a and b, as well as the components of the initial velocity $X^{(1)}$ and $Y^{(1)}$. Other expansion coefficients should be chosen based on the boundary conditions for a particular problem. The convergence of the series is not discussed, but it is assumed that they will be convergent at sufficiently short expansion times (4.43).

Substituting Eq. (4.43) into Eqs. (4.39) and (4.42) yields

$$X_a^{(1)} + Y_b^{(1)} = 0; \tag{4.44}$$

$$X_a^{(2)} + Y_b^{(2)} = -(X_a^{(1)}Y_b^{(1)} - X_b^{(1)}Y_a^{(1)}); \tag{4.45}$$

$$X_b^{(2)} - Y_a^{(2)} = 0. \tag{4.46}$$

We assume that $X^{(1)} = Y^{(1)} = 0$. Then, Eq. (4.44) is satisfied by itself, and from Eqs. (4.45) and (4.46), it follows that the functions $X^{(2)}$ and $Y^{(2)}$ satisfy the Cauchy–Riemann conditions

$$X_a^{(2)} = -Y_b^{(2)} \quad \text{and} \quad X_b^{(2)} = Y_a^{(2)}, \tag{4.47}$$

or are conjugate harmonic functions.

To determine these functions, Eq. (4.47 should be supplemented with boundary conditions. The particles at the bottom should obviously stay at the bottom. It follows that

$$Y(a, 0, t) = 0, \quad 0 < a < \infty; \quad t > 0.$$

Based on this condition, we require that

$$Y^{(2)}(a, 0) = 0 \quad \text{for } 0 \leq a < \infty.$$

We now use our theory to simulate the dam destruction problem. Assume that the dam is vertical (in the Lagrangian plane, it corresponds to $a = 0$), and the fluid occupies a half-band of height $h(0 \leq a \leq \infty; 0 \leq b \leq h)$: see Fig. 4.1. Initial conditions of the problem

$$X(a, b; 0) = a, \quad Y(a, b; 0) = b, \quad X_t(a, b; 0) = 0,$$

$$\text{and} \quad Y_t(a, b; 0) = 0$$

correspond exactly to the original theoretical assumptions. When a dam breaks, the pressure along it (along $a = 0$) suddenly changes

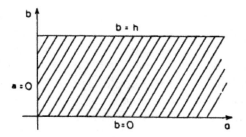

Fig. 4.1. Lagrangian flow plane.

from hydrostatic to zero (on a free surface). The boundary conditions for the pressure will be written as follows:

$$p(a, h; t) = 0, \quad 0 \leq a < \infty, \quad t > 0; \tag{4.48}$$

$$p(0, b; t) = 0, \quad 0 \leq b \leq h, \quad t > 0. \tag{4.49}$$

For free surface particles with $b = h$, according to Eq. (4.48), $p_a = 0$, and based on Eqs. (4.40) and (4.43), we conclude that

$$X^{(2)}(a, h) = 0. \tag{4.50}$$

For free surface particles with $a = 0$, according to Eq. (4.49), the pressure $p_b = 0$, and based on Eqs. (4.41) and (4.43), we find

$$Y^{(2)}(0, b) = -\frac{g}{2}. \tag{4.51}$$

The particles at the bottom should remain there. Hence, we find that

$$Y(a, 0; t) = 0, \quad 0 \leq a < \infty, \quad t > 0.$$

The relation

$$Y^{(2)}(a, 0) = 0, \quad 0 \leq a < \infty \tag{4.52}$$

will be a consequence of this condition.

As was proved, $Z(z) = Y^{(2)} + iX^{(2)}$ is an analytical function of the variable $z = a + ib$ in the half-band shown in Fig. 4.2. On its sides the real or imaginary parts of this function are known (see Eqs. (4.50)–(4.52)). Thus, this function can be determined, e.g., by the conformal mapping method.

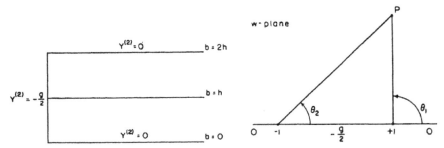

Fig. 4.2. Half-band area with $0 \leq b \leq 2h$ is on the left. The w-plane is on the right.

Since Eq. 4.50 is valid, the relation $X_a^{(2)}(a,h) = 0$ is also valid, and hence, by virtue of Eq. (4.47), the equality $Y_b^{(2)}(a,h) = 0$ is also true. Thus, the harmonic function $Y^{(2)}(a,b)$ can be analytically continued through the line $b = h$ by the method of reflection in the half-band $2h$ (Fig. 4.2, left). The boundary values of $Y^{(2)}$ are shown in the figure. We map this half-band on the upper half-plane using the transform $w = ch(\pi z/2h)$. The vertices $z = 0$ and $z = 2ih$ of the half-band are mapped on points $w = \pm 1$ (see Fig. 4.2, right). The corresponding boundary values of the function $Y^{(2)}$ are also shown in the same figure.

The solution for $Y^{(2)}(w)$ is given by the formula

$$Y^{(2)}(P) = -\frac{g}{2\pi}(\theta_2 - \theta_1),$$

where θ_1, θ_2 are the angles marked in Fig. 4.2 on the right. The analytical function, for which $Y^{(2)}$ serves as the real part, has the form

$$Y^{(2)} + iX^{(2)} = -\frac{ig}{2\pi} \ln \frac{w-1}{w+1}.$$

Moving back to the z-plane, we determine

$$Z(z) = Y^{(2)} + iX^{(2)} = -\frac{ig}{2\pi} \ln \left\{ \frac{ch \frac{\pi z}{2h} - 1}{ch \frac{\pi z}{2h} + 1} \right\}.$$

After separating the real and imaginary parts, we obtain the final result in the form

$$X^{(2)}(a,b) = -\frac{g}{2\pi} \ln \left\{ \frac{\cos^2 \frac{\pi b}{2h} + sh^2 \frac{\pi a}{4h}}{\sin^2 \frac{\pi b}{2h} + sh^2 \frac{\pi a}{4h}} \right\}; \qquad (4.53)$$

$$Y^{(2)}(a,b) = -\frac{g}{\pi} \arctan \left\{ \frac{\sin \frac{\pi b}{2h}}{sh \frac{\pi a}{2h}} \right\}. \qquad (4.54)$$

Due to these relations, the boundary conditions (4.50)–(4.52) can be easily verified.

The shape of the free surface is determined by the expressions

$$X = a + X^{(2)} t^2 \quad \text{and} \quad Y = a + Y^{(2)} t^2, \qquad (4.55)$$

which for $a = 0$ describe the leading edge of moving water and for $b = h$ describe the upper part of its boundary. The results of simulations using Eqs. (4.53)–(4.55) for a dam with a height of 200 ft are shown in Fig. 4.3.

Note that at the point $a = 0, b = 0$, the function $X^{(2)}$ has a logarithmic singularity. Since $p_a^{(0)} = -2\rho X^{(2)}$, the pressure is also singular at this point. This circumstance indicates that the model of an ideal fluid spreading in a potential way is no longer valid near the "spout" of the moving front, and in this region, it is necessary to take into account the viscosity effect (in fact, the wavefront, starting from a certain moment, will collapse, and the flow will become turbulent).

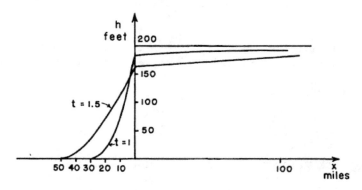

Fig. 4.3. Picture of the dam failure [23].

References

[1] Stokes, G.G. (1847). On the theory of oscillatory waves. *Camb. Trans.*, 8, pp. 441–473.

[2] Whitham, G.B. (1974). *Linear and Nonlinear Waves* (A Wiley-Interscience Publication John Wiley & Sons, New York).

[3] Abrashkin, A.A. and Zen'kovich, D.A. (1990). Vortical steady waves on a shear flow. *Izv. RAN. Fiz. Atm. Okeana*, 26(1), pp. 35–45 [in Russian].

[4] Kochin, N.E., Kibel', I.A. and Roze, N.V. (1964). *Theoretical Hydromechanics*, Vol. 1 (Interscience Publ., New York).

[5] Constantin, A. (2006). The trajectories of particles in Stokes waves. *Invent. math.*, 166, pp. 523–535.

[6] Constantin, A. and Villari, G. (2008). Particle trajectories in linear water waves. *J. Math. Fluid Mech.*, 10, pp. 1–18.

[7] Constantin, A. Ehrnström, M. and Villari, G. (2008). Particle trajectories in linear deep-water waves. *Nonlinear Analysis: Real World Applications*, 9, pp. 1336–1344.

[8] Miche, M. (1944). Mouvements ondulatories de la mer en profondeur constant ou décroissante. *Ann. Ponts Chaussées*, 114, pp. 25–78, 131–164, 270–292, 396–406. (English translation: (1954). *Univ. Calif. Wave Res. Lab.* 3, 363).

[9] Chang, H.K., Liou, J.C. and Su, M.Y. (2007). Particle trajectory and mass transport of finite-amplitude waves in water of uniform depth. *Eur. J. Mech. B/Fluids*, 26, pp. 385–403.

[10] Clamond, D. (2007). On the Lagrangian description of steady surface gravity waves. *J. Fluid Mech.*, 589, pp. 434–454.

[11] Buldakov, E.V., Taylor, P.H. and Taylor, R. Eatock. (2006). New asymptotic description of nonlinear water waves in Lagrangian coordinates. *J. Fluid Mech.*, 562, pp. 431–444.

[12] Nayfen, A.H. (1981). *Introduction to Perturbation Techniques* (John Wiley & Sons, New York).

[13] Zakharov, V.E. (1968). Stability of periodic waves of finite amplitude on the surface of a deep fluid. *J. Appl. Mech. Tech. Phys.*, 9, pp. 190–194.

[14] Hasimoto, H. and Ono, H. (1972). Nonlinear modulation of gravity waves. *J. Phys. Soc. Jpn.*, 33, pp. 805–811.

[15] Davey, A. (1972). The propagation of a weak nonlinear wave, *J. Fluid Mech.*, 53, pp. 769–781.

[16] Yuen, H.C. and Lake, B.M.(1975). Nonlinear deep water waves: Theory and experiment. *Phys. Fluids*, 18, pp. 956–960.

[17] Kuznetsov, E.A. (1977). Solitons in a parametrically unstable plasma. *Sov. Phys. Dokl.*, 22, pp. 507–509.

[18] Ma, Y.-C. (1979). The perturbed plane-wave solutions of the cubic Schrödinger equation, *Appl. Math.*, 60, pp. 43–58.

[19] Peregrine, D.H. (1983). Water waves, nonlinear Schrödinger equations and their solutions. *J. Australian Math. Soc., Ser. B*, 25, pp. 16–43.

[20] Akhmediev, N.N., Eleonskii, V.M. and Kulagin, N.E. (1985). Generation of periodic trains of picosecond pulses in an optical fiber: Exact solutions. *Soviet Phys. JETP*, 62(5), pp. 894–899.

[21] Yuen, H.C. and Lake, B.M. (1982). *Nonlinear Dynamics of Deepwater Waves* (Academic Press, New York, London, Paris).

[22] Lamb, H. *Hydrodynamics*, 6th edn., (1932) (Cambridge University Press, Cambridge).

[23] Stoker, J.J. (1957). *Water Waves: The Mathematical Theory with Applications* (Interscience Publishers, New York, London).

[24] Pohle, F.V. (1950). *The Lagrangian Equations of Hydrodynamics: Solutions which are Analytic Functions of the Time*, Thesis (New York University).

[25] Pohle, F.V. (1952). *Motions of Water due to Breaking of a Dam, and Related Problems*. U. S. National Bureau of Standards, Gravity Waves, N. B. S. Circular 521.

Chapter 5

Gerstner Wave

> "The Gerstner wave is the first exact solution in
> nonlinear wave theory."

Commonly, the origin of nonlinear wave science is associated with the first experiments conducted by Scott Russell, who in the 1830s and 1840s first observed solitons running along the surface of a shallow channel [1]. In 1895, Korteweg and de Vries gave a mathematical description of this phenomenon based on the equation later named after them [2]. However, for many researchers, the first analytical description of a nonlinear wave, published by Franz Joseph Gerstner in 1802, remained (and remains) out of sight [3, 4]. Due to a number of circumstances, the Gerstner wave was not given even "a thousandth" of the attention that was given to the Korteweg–de Vries solitons. But the fact remains: In the beginning, there was the Gerstner wave.

Brief biography. The name Franz Josef Gerstner was well known in the Czech Republic and Austria (see Fig. 5.1). He was a professor and director of the Polytechnic School in Prague and head of the construction of the entire water supply system. Virtually no engineering company in the Czech Republic could do without his participation or advice [5]. Hans Straub (1895–1964), director of the Higher Technical School in Zurich, included F.J. Gerstner in the list of names of "great researchers and engineers" [6], not only as the scientist who proposed the construction of the railway but also as the creator of the theory of trochoidal waves. It is interesting to add that his son

Fig. 5.1. Prague engineer and mechanic Franz Josef Gerstner (1756–1832).

(also a very famous engineer), Franz Anton Gerstner, was the builder of the first railway in Russia.

The Korteweg–de Vries equation is derived in the approximation of small nonlinearity and dispersion from the complete equations of fluid dynamics. Its soliton solution is a classic example of a nonlinear wave in which the effects of nonlinearity and dispersion counterbalance each other. On the other hand, the Gerstner wave represents an exact solution of the complete equations of fluid dynamics and is so far the only example of their integration for gravity waves in deep water. It is difficult to deny that, from a mathematical point of view, the Gerstner wave is a more significant achievement of the analytical theory of nonlinear waves on water than solitons in shallow water.

5.1. Basic Properties

Consider the plane motions of an ideal incompressible fluid with a free surface in the Earth's gravity field. Let the X-axis be horizontal, directed to the right, and the Y-axis directed vertically upward. Similarly, let a be the Lagrangian horizontal coordinate and b the vertical one (the fluid corresponds to the half-plane $b \leq 0$). We consider the fluid to be infinitely deep. The Lagrange equations of fluid dynamics can be written in the form

$$\frac{\partial}{\partial t} \frac{D(X,Y)}{D(a,b)} = \frac{\partial D_0}{\partial t} = 0, \tag{5.1}$$

$$\frac{\partial}{\partial t}(X_{ta}X_b + Y_{ta}Y - X_{tb}X_a - Y_{tb}Y_a) = \frac{\partial(\Omega D_0)}{\partial t} = D_0\frac{\partial \Omega}{\partial t} = 0. \tag{5.2}$$

The second of these equations reflects the condition for the conservation of vorticity Ω. To describe a wave on the surface of a fluid, it is necessary to find a solution to Eqs. (5.1) and (5.2) and also satisfy the condition of constant pressure on the free surface,

$$p|_{b=0} = p_0 = \text{const}, \tag{5.3}$$

and the condition of no leakage condition at the bottom, $Y_t|_{b\to-\infty} = 0$. Gerstner indicated the exact solution to this problem [3,7–10]:

$$X = a - Ae^{kb}\sin(ka - \omega t), \quad Y = b + Ae^{kb}\cos(ka - \omega t),$$
$$\text{and} \quad b \leq 0, \tag{5.4}$$

where A is the wave amplitude, k is the wave number, and ω is the wave frequency. As in linear potential waves, these values are connected by the relation

$$\omega^2 = gk, \tag{5.5}$$

which is a consequence of condition (5.3). At each moment of time, the free surface of the wave is a trochoid, i.e., a curve drawn by some point of a circle of radius A, rolling without sliding along a horizontal line $Y = -A$. Remaining unchanged, the trochoid moves at a speed $c_0 = \omega k^{-1}$ to the right. Due to this, Gerstner waves are also called trochoidal. Only solutions with $A \leq k^{-1}$ have a physical meaning; otherwise, the profile is self-intersecting. If $A = k^{-1}$, then the wave crests become sharp (the angle of sharpening is zero); such a limiting trochoid is called a cycloid.

In the 1860s, the solution to Eq. (5.4) was rediscovered by three authors at once, namely, Froude [11], Rankine [12], and Reech [13]. For more than half a century, Gerstner's classical result remained

94   *Analytical Fluid Dynamics in Lagrangian Variables*

unnoticed, and this circumstance characterizes Gerstner as an outstanding scientist who was ahead of his time.

The coordinates of the trajectory of an individual particle satisfy the relation (see Eq. (5.4))

$$(X - a)^2 + (Y - b)^2 = A^2 e^{2kb},$$

from which it follows that in a fixed reference frame, each particle moves along a circle of radius Ae^{kb} (there is no particle drift in Gerstner waves). The initial position of the fluid particles in the Gerstner solution does not coincide with the Lagrangian coordinates, namely,

$$X_0 = a - Ae^{kb} \sin ka \quad \text{and} \quad Y_0 = b + Ae^{kb} \cos ka.$$

Of course, one can take X_0 and Y_0 as new Lagrangian variables, but in this case, the expressions for X and Y can no longer be written explicitly as functions X_0 and Y_0.

The vorticity of Gerstner waves is equal to

$$\Omega_* = \frac{2k^3 A^2 c_0 \cdot \exp(2kb)}{1 - k^2 A^2 \cdot \exp(2kb)}. \tag{5.6}$$

To calculate the vorticity, we use Eqs. (5.2) and (5.4). In the case of small-wave steepness, where $\varepsilon = kA \ll 1$, the vorticity is represented by the expression

$$\Omega_* = 2kc_0 e^{2kb} \varepsilon^2 (1 + e^{2kb} \varepsilon^2) + O(\varepsilon^6). \tag{5.7}$$

It shows that in the linear approximation, the Gerstner wave is vortex free and coincides with the linear Stokes wave. In the quadratic approximation, the vorticity of the Gerstner wave is $2kc_0 e^{2kb} \varepsilon^2$. The vorticity is equal in absolute value and opposite in sign to the Stokes drift vorticity, which is equal to $-U_{sb}$ (see Sec. 4.1.3). Comparing Eqs. (5.4) and (4.16) and 4.17, we can conclude that in the quadratic approximation, the following formula is valid [14]:

$$\text{Stokes Wave} = \text{Gerstner wave} + \text{Stokes drift.}$$

Recall that the term $\varepsilon^2 e^{2kb}/2k$ ensures a zero average level of the fluid. When comparing the Stokes and Gerstner waves, this term should be discarded, since the average level is not preserved for the

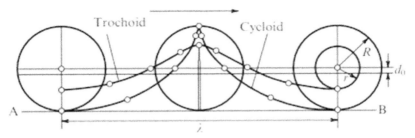

Fig. 5.2. Cycloid ($R = k^{-1} = (2\pi)^{-1}\lambda$) and trochoid ($r < R$). The average level of the cycloid is located below the average level of the trochoid by the magnitude $d_0 = \pi(R^2 - r^2)\lambda^{-1}$. In terms of the Gerstner wave (5.4), this means that the greater the amplitude A, the lower the equilibrium level of the fluid [7].

Gerstner wave (see the caption to Fig. 5.2). We repeat that Stokes did not impose this condition and wrote down his solution without this term [15]. It is interesting to add that the profiles of the Stokes and Gerstner waves coincide up to a cubic approximation in the small steepness parameter [16]. The difference between them is observed only in the fourth order of perturbation theory.

Using the solution (5.4) and Eqs. (4.40) and (4.41), we determine the pressure [7]:

$$\frac{p - p_0}{\rho} = -gb - \frac{\omega^2 A^2}{2}(1 - e^{2kb}). \tag{5.8}$$

The pressure depends only on the b coordinate. As shown by A.S. Monin [17], the only type of steady-state wave in which the pressure depends only on the vertical Lagrangian coordinate is the Gerstner trochoidal wave. It is remarkable that for a stratified fluid with density $\rho(b)$, solution (5.4) also remains valid [18]. This is due to the fact that both density and pressure depend on only one coordinate b, which means they are functions of each other. For a barotropic fluid, the condition for conservation vorticity (5.2) is the same as for a homogeneous fluid.

5.1.1. *Implementation problem*

In the theory of waves on water, the Stokes wave has become more famous and attracted significantly more attention than the Gerstner wave. This seems surprising since the Stokes solution is written as a series with respect to a small parameter of wave steepness. Such a representation for a periodic surface wave turned out to be more practically important than an exact solution. The reason for this is the problem of implementing the Gerstner wave. Unlike the Stokes wave, it is a vortex wave and cannot arise in nature under the action of potential forces (Lagrange theorem). To obtain the Gerstner wave, either external non-potential forces or special initial conditions are necessary. Thus, H. Lamb suggested that the Gerstner wave can be born from a shear flow that has the same vorticity as the wave [4]. In this case, the translational motion of fluid particles in the flow should be transformed into rotation in a circle (there is no drift in the Gerstner wave). The specificity of such a scenario clearly did not contribute to the popularization of the Gerstner solution and its widespread use for practical calculations. Nevertheless, based on Gerstner's formulas, A.N. Krylov developed the theory of ships rolling on undulation [19], "which has found great application in shipbuilding" [16].

The discovery of the mechanism of modulation instability for potential waves on water [20] would seem to have finally turned Gerstner's theory into a fluid dynamics artifact, i.e., a beautiful exact solution, impossible to implement in nature, but suddenly, at approximately the same time, examples of its new applications began to appear. Pollard modified the Gerstner solution for waves in a rotating fluid in the f-plane approximation [21]. Yih [22], Mollo-Christensen [23, 24], and, in a more complete form, Constantin [25] applied Gerstner's solution to describe edge waves propagating along a sloping beach. In addition, Mollo-Christensen described the billows of Gerstner waves (trochoidal clouds) in a stratified atmosphere [26].

All these achievements (we will talk about them in the following) have significantly increased the status of Gerstner's solution, but the question of its physical implementation still remained open. However, this long-standing problem was solved by Monismith *et al.* [27], who succeeded in creating the Gerstner wave in laboratory conditions. The principle of the experiment was as follows. In the Gerstner wave, fluid particles move in a circle, so there is no drift stream there. This distinguishes it from a Stokes wave, the propagation of which is accompanied by particle transport (Stokes drift). This drift is directed along the motion of the wave; therefore, when the Gerstner wave was generated, a current moving in the opposite direction was created. The absence of drift of fluid particles testified in favor of observing the Gerstner wave in the experiment. It is necessary to emphasize the particular care with which Monismith *et al.* [27] formulated their conclusions. To further verify them, they addressed similar experiments carried out in other laboratories and showed that Gerstner waves had previously been observed in three more trays [28–30].

All these results were obtained in limited channels with artificially (mechanically) generated waves. But, as Monismith *et al.* [27] noted, similar observations (i.e., the absence of drift of fluid particles) also took place for waves in the open ocean [31]. Thus, the existence of Gerstner waves was also confirmed by field observations. However, with a steepness exceeding 1/3, Gerstner waves are unstable to three-dimensional perturbations [32], but now it has become possible to talk about them as real physical oscillations. In turn, Weber pointed out that when viscosity and surface films are taken into account, a drift steam occurs in the Gerstner wave [33]. In his opinion, such wave oscillations could well have been observed by experimenters in laboratory tanks for a long time (without realizing this fact and without being attached to Gerstner waves).

It is difficult to state whether this is due to the experimental work [27] that inspired the theorists, but in the last decade, Gerstner's topic has gained a second wind. Many papers generalizing Gerstner's solution have appeared in the case of pressure variability on a free surface due to the wind action with allowance for the Earth's rotation and stratification. A list of these papers will be given in the following. The terms Gerstner-like or Gerstner-type appeared in a number of these publications. They reflect the relation between the solutions obtained and the classical Gerstner wave.

5.2. Edge Waves on a Sloping Beach

Edge waves are waves that propagate along the beach. They reach their maximum amplitude at the border with land, and decrease amplitude rapidly when moving away from the beach. All the energy of these waves is concentrated in a narrow coastal area and is not actually transferred to the open ocean; so, as they say, wave energy is "trapped." As a result, the edge waves are also referred to as trapped waves. The study of these waves, as in the case of common surface

98   *Analytical Fluid Dynamics in Lagrangian Variables*

waves, was initiated by Stokes [34]. By now, this is already a separate section of the theory of waves on water (see, e.g., [35, 36]), but the only known exact solution in nonlinear wave theory is again related to the Gerstner approach [22–25]. Let us present it on the basis of another work [25].

Let a sloping bottom approach the horizontal at an angle $\alpha(0 < \alpha < \pi/2)$. We will consider edge waves that propagate along the beach. We choose the X-axis parallel to the direction of the beach, the Y-axis along the sloping bottom, and the Z-axis perpendicular to it (see Fig. 2.2). The water edge satisfies the following conditions: $-\infty < X < \infty, Y = b_0$, and $Z = 0$, where b_0 is a constant value $(b_0 \leq 0)$. The unit mass of a fluid in such a coordinate system will be affected by a force equal to

$$\overrightarrow{f} = (0, -g\sin\,\alpha, -g\cos\,\alpha). \qquad (5.9)$$

In the absence of waves, the fluid will be enclosed between the bottom plane $(Z = 0)$ and the plane $Z = (b_0 - Y)\tan\alpha$. Let us choose the fluid to coincide with the array of Lagrangian coordinates, and the variables a, b, and c will be counted, respectively, along the X, Y, and Z axes. The equation of a free surface in a Lagrangian system has the form

$$c = (b_0 - b)\tan\alpha; \quad b \leq b_0. \qquad (5.10)$$

Consider the motion when fluid particles move in planes parallel to the bottom. In this case, the Z component of the velocity is missing and the no-leakage condition is fulfilled by itself. Compared with Eq. (5.4), we write down the representation of such a two-dimensional flow as follows [25]:

$$\begin{cases} X = a - \dfrac{1}{k}e^{k(b-c)}\sin(ka + \sqrt{gk \cdot \sin\alpha}t), \\[2mm] Y = b - c + \dfrac{1}{k}e^{k(b-c)}\cos(ka + \sqrt{gk \cdot \sin\alpha}t), \\[2mm] Z = c + c\tan\alpha - \dfrac{\tan\alpha}{2k}e^{2kb_0}(1 - e^{-2kc(1+\cot\alpha)}). \end{cases} \qquad (5.11)$$

The first two equations of this system are similar to Gerstner's formulas, only the roles of the b coordinate and the free-fall acceleration

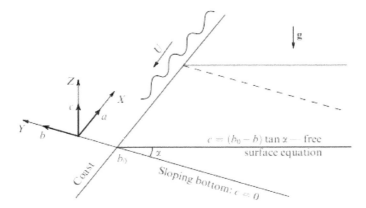

Fig. 5.3. Boundaries of a fluid in a Lagrangian coordinate system.

in them are played by $b-c$ and $g\sin\alpha$. The wave frequency is equal to $\sqrt{gk\cdot\sin\alpha}$ (naturally, the Stokes edge wave has the same frequency). The plus sign in the arguments of trigonometric functions means that the wave runs toward negative X (its velocity $U = \sqrt{g\sin\alpha/k}$, see Fig. 5.3). The profile of the wave on the beach ($Z = c = 0$) corresponds to a trochoid without points of sharpening ($b_0 < 0$) or a cycloid with upward points of sharpening ($b_0 = 0$).

The relation for Z in Eq. (5.11) ensures that the condition of constant pressure on a free surface is fulfilled. Substituting condition (5.10) into Eq. (5.11), we obtain a parametric representation for a free surface. It is easy to see that when moving away from the beach ($b \to -\infty$), the wave amplitude decreases exponentially, i.e., it is trapped. The vorticity of the wave has only a Z component, and its magnitude is obtained from Eq. (5.6) by replacing $A \to k^{-1}\exp(-kc)$ and adding a minus sign ahead (the wave runs to the left).

The instability of three-dimensional edge waves (5.11) was examined by Ionescu-Kruse [37] using the Leblanc short-wave perturbation method [32]. It is proved that waves with a steepness above $\frac{7}{18}\sin\alpha$ are unstable.

5.3. Waves in Layers with Density Discontinuities

Mollo-Christensen [38] drew attention to two original circumstances related to the Gerstner wave.

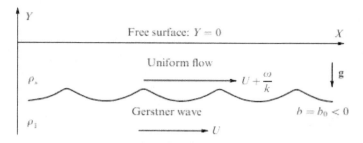

Fig. 5.4. Two-layer Mollo-Christensen model.

Consider a two-layer fluid model (Fig. 5.4). Let the densities of the lower and upper layers be constant and equal to ρ_1 and $\rho_*(\rho_1 > \rho_*)$, respectively. Suppose that the Gerstner wave (5.4) propagates in the lower fluid against the background of a uniform flow U, and the upper one is a uniform flow with a velocity equal to the velocity of the Gerstner wave $U + \omega/k$. In this case, the velocity at the interface will be continuous, and the equality of pressures on both sides is fulfilled if we assume that the frequency of the wave ω satisfies the condition

$$\omega^2 = \frac{\rho_1 - \rho_*}{\rho_1} gk.$$

It coincides with the dispersion relation of linear waves at the interface of two fluids. The solution constructed in this way describes the internal Gerstner wave on a uniform flow and can be generalized both to the case of an arbitrarily stratified bottom fluid and to a uniformly rotating fluid [38].

A special case of a two-layer Mollo-Christensen model, when the upper fluid is stationary and $U = -\omega/k$, was studied [39]. This condition corresponds to a stopped wave, i.e., the motion of fluid particles with density ρ_1 along trochoidal trajectories in the negative direction of the X-axis (the author [39] does not note this circumstance). Mollo-Christensen's ideas were developed in a more complete form in relation to zonally traveling waves in the equatorial region. Internal Gerstner waves [40], as well as their transformations on the current [41] and along with it the meridional flow, are described in the f plane approximation [42].

The second find by Mollo-Christensen [38] is even more beautiful. Let us rewrite Eq. (5.4) by changing the sign before the sine and in

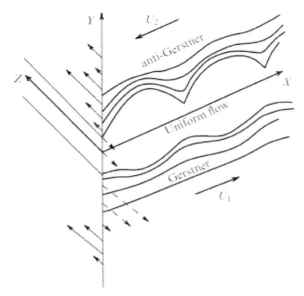

Fig. 5.5. Three-layer Mollo–Christensen model. Flow directions are shown for geostrophic billows [38]. Arrows along the Z-axis show the meridional flow with the velocity profile changing its direction.

the exponents (the flow is studied in the XOY plane):

$$X = a + Ae^{-kb}\sin(ka - \omega t), \quad Y = b + Ae^{-kb}\cos(ka - \omega t). \quad (5.12)$$

These relations also satisfy the system of equations of fluid dynamics (5.1) and (5.2), but they will describe amplitude-limited waves if we assume $b > 0$. The wave amplitude will decrease with an increase in the vertical Lagrangian coordinate, which corresponds to an inverted Gerstner trochoidal wave, but only with its troughs now facing upward. Let us call such waves anti-Gerstner waves.

As a result, a model of three-layer fluid flow was proposed [38] (see Fig. 5.5):

— in the lower fluid ($b = b_1 < 0$) with density ρ_1, the Gerstner wave with frequency ω_1 runs along the X-axis against the background of a uniform flow U_1;
— in the upper fluid ($b = b_2 > 0$) with density ρ_2, an anti-Gerstner wave with frequency ω_2 runs along the X-axis against the background of a uniform flow U_2;

102 *Analytical Fluid Dynamics in Lagrangian Variables*

— in the central part ($b_1 \leq b \leq b_2$), a wave-trapped fluid with density ρ_* ($\rho_1 > \rho_* > \rho_2$) flows uniformly in the direction of the X-axis at a speed

$$U_1 + \frac{\omega_1}{k} = U_2 + \frac{\omega_2}{k},$$

where k is the same wave number for both waves. Expressions for frequencies are determined from the condition of continuous pressure at the interface [38].

This flow diagram serves as a model of clouds trapped by waves and moving with them. Mollo-Christensen used the term "billows" for them. Many cloud billows do not contain condensate and therefore remain invisible, but their characteristic feature is exactly the finite perturbation amplitude. This chapter considers two cases, namely, gravity billows in a non-rotating continuum and geostrophic billows ($g = 0$, but the rotation of the atmosphere is taken into account). The combined case, where both the effect of gravity and the Earth's rotation are considered, is obtained by combining these two solutions. In relation to the ocean, the idea of a three-layer model for the Gerstner–anti-Gerstner system was employed [43].

Weber, studying the properties of weakly nonlinear waves at the boundary of two fluids [44, 45], suggests calling any wave for which there is no drift of fluid particles Gerstner-like. Obviously, all the exact solutions analyzed and mentioned by us have this property. In this sense, it seems expedient to discuss a family of different waves that can be combined under the name Gerstner and called Gerstner-like.

References

[1] Russell, J.S. (1844). Report on Waves. *Rept. 14th meetings of the British Assoc. for the Advancement of Science*, London: John Murray, pp. 311–390.

[2] Korteweg, D.J. and de Vries, G. (1895). On the change of form of long waves advancing in a rectangular channel, and on a new type of long stationary waves. *Phil. Mag.*, 39, pp. 422–443.

[3] Gerstner, F. (1802) Theorie der Wellen. *Abh. d.k. böhm. Ges. d. Wiss.* (Also see: (1809) *Ann. Phys. Lpz.* 2, p. 412).

[4]   Lamb, H. (1932). *Hydrodynamics*, 6th ed. (Cambridge University Press, Cambridge).

[5]   Voronin, M.I. and Voronina, M.M. (1994). *Franz Anton Gerstner* [in Russian], (Nauka, St. Petersburg).

[6]   Straub, H. (1992). *Die Geschichte der Bauingenieurkunst* (Springer, Basel).

[7]   Kochin, N.E. Kibel, I.A. and Roze, N.V. (1964). *Theoretical Hydromechanics*, Vol. 1 (Interscience Publ., New York).

[8]   Constantin, A. (2011). Nonlinear water waves with applications to wave–current interactions and tsunamis. *CBMS-NSF Conference Series in Applied Mathematics*, Vol. 81 (PA: SIAM, Philadelphia).

[9]   Constantin, A. (2001). On the deep water wave motion. *J. Phys. A Maths. Gen.*, 34, pp. 1405–1417.

[10] Henry, D. (2008). On Gerstner's water wave. *J. Nonlinear Math. Phys.*, 15, pp. 87–95.

[11] Froude, W. (1862). On the rolling of ships. *Trans. Inst. Naval. Arch.*, 3, pp. 45–62.

[12] Rankine, W.J.M. (1863). On the exact form of waves near the surface of deep water. *Phil. Trans. R. Soc. Lond.* A, 153, pp. 127–138.

[13] Reech, F. (1869). Sur la thèorie des ondes liquids pèriodiques. *C. R. Acad. Sci. Paris*, 68, pp. 1099–1101.

[14] Abrashkin, A.A. and Pelinovsky, E.N. (2018). On the relation between Stokes drift and Gerstner wave. *Phys.-Usp.*, 61(3), pp. 307–312.

[15] Stokes, G.G. (1847) On the theory of oscillatory waves. *Cambridge Trans.*, 8, pp. 441–473; Stokes, G.G. (1880). *Mathematical and Physical Papers*, Vol. 1, pp. 197–229 (Cambridge University Press).

[16] Sretensky, L.N. (1977). *Theory of Wave Motions of a Fluid* [in Russian] (Nauka, Moscow).

[17] Monin, A.S. (1972). Lagrangian Description of Steady-state Waves. *Dokl. Akad. Nauk SSSR*, 203, 4, pp. 769–771 [in Russian].

[18] Dubreil-Jacotin, M.L. (1932). Sur les ondes de type permanent dans les liquids heterogenes. *Atti. Accad. Lincei. Rend. Cl. Sci. Fis. Mat. Nat.*, 6(15), pp. 814–819.

[19] Kriloff, A.N. (1898). A general theory of the oscillations of a ship waves, *Trans. Inst. Naval. Archit.*, 40, pp. 135–190.

[20] Benjamin, T.B. and Feir, J.E. (1967). The disintegration of wavetrains on deep water. Part 1. Theory. *J. Fluid Mech.*, 27, pp. 417–430.

[21] Pollard, R.T. (1970). Surface waves with rotation: An exact solution. *J. Geophys. Res.*, 75, pp. 5895–5898.

[22] Yih, C.-S. (1966). Note on edge waves in a stratified fluid. *J. Fluid Mech.*, 24, pp. 765–767.

[23] Mollo-Christensen, E. (1982). Allowable discontinuities in a Gerstner wave. *Phys. Fluids*, 25, pp. 586–587.

[24] Mollo-Christensen, E. (1978). Edge waves in a rotating stratified fluid, an exact solution. *J. Phys. Ocean.*, 9, pp. 226–229.

[25] Constantin, A. (2001). Edge waves along a sloping beach. *J. Phys. A Maths. Gen.*, 34, pp. 9723–9731.

[26] Mollo-Christensen, E. (1978). Gravitational and geostrophic billows: some exact solutions. *J. Atm. Sci.*, 35, pp. 1395–1398.

[27] Monismith, S.G., Cowen, E.A., Nepf, H.M., Magnaudet, J. and Thais, L. (2007). Laboratory observations of mean flows under surface gravity waves. *J. Fluid Mech.*, 573, pp. 131–147.

[28] Swan, C. (1990). Experimental study of waves on a strongly sheared current profile. *In Proc. 22nd Intl Coastal Engng Conf.*, pp. 489–502.

[29] Jiang, J.Y. and Street, R.L.S. (1991). Modulated flows beneath wind-ruffled, mechanically generated waves. *J. Geophys. Res.* (Oceans) 96, pp. 2711–2721.

[30] Thais, L. (1994). Contribution a l'ètude du movement turbulent sous des vagues de surface cisaillèes par le vent. *Thèse Inst. Nat. Polytech. De Toulouse* (Toulouse).

[31] Smith, J.A. (2006). Observed variability of ocean wave Stokes drift, and the Eulerian response to passing groups. *J. Phys. Oceanogr.*, 36(7), pp. 1381–1402.

[32] Leblanc, S. (2004). Local stability of Gerstner's waves *J. Fluid Mech.*, 506, pp. 245–254.

[33] Weber, J.E.H. (2011). Do we observe Gerstner waves in wave tank experiments? *Wave Motion*, 48, pp. 301–309.

[34] Stokes, G.G. (1846). Report on recent researches in hydrodynamics. *Rep.* 16th *Brit. Assoc. Adv. Sci.*, pp. 1–20; see also: Stokes, G.G. (1880). *Papers*, Vol.1, pp. 157–187 (Cambridge University Press).

[35] Johnson, R.S. (2007). Edge waves: Theories past and present. *Phil. Trans. R. Soc.* A, 365, pp. 2359–2376.

[36] Dubinina, V.A., Kurkin, A.A., Pelinovsky, E.N. and Polukhina, O.E. (2004). *Izvestya, Atmospheric and Oceanic Physics*, 40(4), pp. 464–470.

[37] Ionescu-Kruse, D. (2014). Instability of edge waves along a sloping beach, *J. Differ. Equations*, 256, pp. 3999–4012.

[38] Mollo-Christensen, E. (1978). Gravitational and geostrophic billows: Some exact solutions. *J. Atmos. Sci.*, 35, pp. 1395–1398.

[39] Stuhlmeier, R. Internal Gerstner waves: Applications to dead water. *Appl. Anal.*, 93(7), pp. 1451–1457.

[40] Hsu, H.-C. (2014). An exact solution for nonlinear internal equatorial waves in the f-plane approximation, *J. Math. Fluid Mech.*, 16, pp. 463–471.

[41] Henry, D. (2015). Internal equatorial water waves in the f-plane. *J. Nonlinear Math. Phys.*, 22, pp. 499–506.

[42] Rodriguez-Sanjurjo, A. (2018). Internal equatorial water waves and wave-current interactions in the f-plane. *Monatsh. Math.*, 186, pp. 685–701.

[43] Stuhlmeier, R. and Stiassnie, M. (2014). Progressive waves on a blunt interface. *Discr. Contin. Dyn. Syst.*, 34(8), pp. 3171–3182.

[44] Weber, J.E.H. (2018). An interfacial Gerstner-type trapped wave. *Wave Motion*, 77, pp. 186–194.

[45] Weber, J.E.H. (2019). A Lagrangian study of internal Gerstner- and Stokes type gravity waves. *Wave Motion*, 88, pp. 257–264.

Chapter 6

Weakly Vortex Waves

6.1. Guyon Waves on Deep Water

"You can't stop the waves,
but you can learn to surf."

— Jon Kabat-Zinn

In the theory of nonlinear waves in deep water in the absence of currents, two model solutions, namely, the Stokes wave and the Gerstner wave, appear. Their properties differ significantly. The Stokes wave is a potential wave and the Gerstner wave is a vortex one. There is a drift of fluid particles in the Stokes wave, but not in the Gerstner wave. In the Stokes wave, fluid particles move in the direction of wave propagation, while in the Gerstner wave there is no such effect: Fluid particles rotate in a circle. For weakly nonlinear Gerstner waves, unlike the Stokes wave, there is no modulation instability effect. If we try to write the nonlinear Schrödinger equation (NLSE) for a train of Gerstner waves, the coefficient for cubic nonlinearity will be zero [1, 2].

However, there is a wave model that occupies an intermediate position between the two mentioned types of waves. We are talking about the class of waves that Guyon proposed for consideration [3,4]. These are nonlinear stationary waves in a weakly vortical fluid. They are distinguished from the Stokes wave by the presence of vorticity and from the Gerstner wave by the existence of fluid particle drift. In contrast to Gerstner waves, Guyon waves are introduced

107

as oscillations with a fairly general distribution of vorticity Ω. The latter is specified in the following form:

$$\Omega = \sum_{n=1}^{\infty} \varepsilon^n \cdot \Omega_n(\psi). \tag{6.1}$$

Here, ε is a small parameter of wave steepness, Ω_n is the vorticity of approximation n, and ψ is the stream function. This formula is the most general representation of the vorticity of a stationary flow in the absence of a shear flow (vortex of zero approximation). Guyon proposed a general scheme of perturbation theory for periodic waves with a vorticity distribution (6.1), proved the convergence of this scheme, and found explicit solutions for the first two approximations [3, 4]. For this purpose, an original approach was developed to describe flows in variables (x, ψ), where x is the Cartesian Eulerian coordinate. Consider the properties of Guyon waves based on the method of modified Lagrangian coordinates, which permits one to compare Guyon waves with both the Gerstner wave and the Stokes wave [5]. The vorticity in our description is specified as follows:

$$\Omega = \sum_{n=1}^{\infty} \varepsilon^n \cdot \Omega_n(b), \tag{6.2}$$

where b is the vertical Lagrangian coordinate. The pattern of streamlines in the reference frame associated with the wave is shown in Fig. 6.1 (the first two terms of the series (6.2) are written down). Within the framework of the proposed approach, weakly nonlinear Gerstner waves and Stokes waves will be obtained as special cases of Guyon waves by choosing expressions for vortices Ω_n.

The formulation of the problem for the Guyon wave exactly corresponds to the content, notation, and formulas provided in Section (4.1) with one exception. Since the vorticity of the wave disturbances is now non-zero, Eq. (4.12) can be written as follows:

$$\Omega = -\sigma' + \overline{\frac{D(\sigma\xi_q, \xi)}{D(q, b)}} + \overline{\frac{D(\sigma\eta_q, \eta)}{D(q, b)}} = \Omega(b). \tag{6.3}$$

We recall that the bar in this formula is the sign of averaging over the variable q at the wavelength.

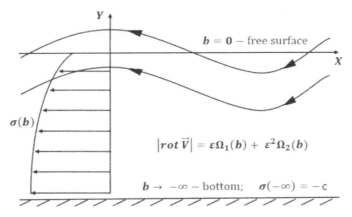

Fig. 6.1. Picture of a stationary flow in a wave-related reference frame: periodic flow perturbations with a $\sigma(b)$ profile.

We will be interested in the first three approximations. Within the framework of such a description, the vorticity is determined by the relation

$$\Omega = \varepsilon\Omega_1(b) + \varepsilon^2\Omega_2(b) + \varepsilon^3\Omega_3(b) + O(\varepsilon^4), \qquad (6.4)$$

where $\Omega_1(b), \Omega_2(b)$, and $\Omega_3(b)$ are given functions. It is these functions that will determine the specifics of the propagation of the vortex waves being studied

6.1.1. *Linear approximation*

We write the equations of the first approximation, according to Eqs. (4.4)–(4.6), in the form

$$\xi_{1q} + \eta_{1b} = 0, \qquad (6.5)$$

$$\sigma_0^2 \xi_{1qq} = -H_{1q}, \qquad (6.6)$$

$$\sigma_0^2 \eta_{1qq} = -H_{1b}. \qquad (6.7)$$

We differentiate Eq. (6.6) by q, Eq. (6.7) by b, and add them up. With allowance for Eq. (6.5), we obtain

$$H_{1qq} + H_{1bb} = 0.$$

We now choose a periodic solution of this equation with period λ in the variable q:

$$H_1 = H_1^* e^{kb} \cos kq, \qquad (6.8)$$

110   *Analytical Fluid Dynamics in Lagrangian Variables*

where H_1^* is the constant. Substituting this expression into Eqs. (6.6) and (6.7), we determine

$$\xi_{1_{qq}} = \frac{kH_1^*}{\sigma_0^2}e^{kb}\sin kq; \quad \eta_{1_{qq}} = -\frac{kH_1^*}{\sigma_0^2}e^{kb}\cos kq. \qquad (6.9)$$

For simplicity, it is convenient to assume that $H_1^* = \sigma_0^2$. By integrating Eq. (6.9) and taking into account the periodicity of perturbations ξ_1 and η_1 with respect to the variable q, we find

$$\xi_1 = -\frac{1}{k}e^{kb}\sin kq + l_1(b); \quad \eta_1 = \frac{1}{k}e^{kb}\cos kq + m_1(b). \qquad (6.10)$$

By substituting these relations into the continuity equation of this approximation (6.5), we obtain

$$\frac{\partial m_1(b)}{\partial b} = 0,$$

which implies that $m_1 = $ const. To ensure that the average level coincides with the horizontal $Y = 0$, i.e., the condition

$$\int_0^\lambda \eta_1 dq = 0$$

is fulfilled (see Eq. (4.9)), the constant m_1 should be assumed to be zero. We will also choose the function $l_1(b)$ equal to zero, but for completely different reasons.

Our description of the flows is that they can be considered as a mapping of a certain region of variables q and b on the plane of variables X and Y. If $l_1 = 0$, the half-band $\{0 \le q \le \lambda, b \le 0\}$ is mapped on the region $\{0 \le X \le \lambda, Y \le \eta_1(q, 0)\}$. If $l_1 \ne 0$, the mapping area will be shifted by $l_1(b)$, i.e., by a different value on each Lagrangian horizon $b = $ const in the general case. The type of function l_1 does not affect the flow velocity, waveform, and other characteristics, so for simplicity of calculations, we will consider it equal to zero. In the Lagrangian description, the particles of a fluid can be renumbered in countless ways, but this, obviously, should not affect the mathematical representation of the flow in any way. Our choice of mapping between Lagrangian and physical regions with $l_1 = 0$ is thus dictated by considerations of convenience of description.

Since in the flow region $b \leq 0$ the oscillation velocities of the fluid particles $\sigma_0 \xi_{1_q}$ and $\sigma_0 \eta_{1_q}$ decrease exponentially with the depth, the condition (4.7) is fulfilled. For pressure to be constant, the following equality should be fulfilled (see Eq. (4.8)):

$$H_1(q,0) = g\eta_1(q,0),$$

whence it follows that $\sigma_0^2 = g/k = c_0^2$. The propagation velocity c_0 of the wave coincides with the phase velocity of linear potential waves. The frequency of the wave is equal to $\sqrt{gk}$, as for a linear potential wave.

The expression for a linear Guyon wave in Lagrangian variables a and b can be written as follows:

$$X = a - c_0 t + \varepsilon \sigma_1(b)t - \varepsilon \cdot \frac{1}{k}e^{kb} \sin k(a - c_0 t);$$

$$Y = b + \varepsilon \cdot \frac{1}{k}e^{kb} \cos k(a - c_0 t). \tag{6.11}$$

It was taken into account that $\sigma_0 = -c_0$ (see Eq. (4.14)). In addition to oscillatory motion, fluid particles participate in the non-uniform drift, which is related to the $\sigma_1(b)$ function. It follows from Eq. (6.3) that it is related to the vorticity of the first approximation:

$$\sigma_1' = -\Omega_1.$$

This function can conveniently be represented as

$$\sigma_1(b) = \sigma_1(-\infty) - \int_{-\infty}^{b} \Omega_1(b)db = -c_1 - \int_{-\infty}^{b} \Omega_1(b)db. \tag{6.12}$$

It is seen that specifying the vorticity of $\Omega_1(b)$ is not enough to fully define the function $\sigma_1(b)$. In the first approximation, it is found up to a constant, namely, a linear correction to the propagation velocity c_0, taken with a minus sign. Its value should be calculated in the next approximation. This feature of specifying σ_n is inherent in all other approximations. The integral term in Eq. (6.12) corresponds to the drift velocity of fluid particles,

$$u_1(b) = \sigma_1(b) + c_1 = -\int_{-\infty}^{b} \Omega_1(b)db.$$

112 *Analytical Fluid Dynamics in Lagrangian Variables*

Table 6.1. Properties of the considered triad of waves in the linear approximation.

Name/Properties	Stokes Wave	Gerstner Wave	Guyon Wave
Vorticity	$= 0$	$= 0$	$\neq 0$
Drift stream	$= 0$	$= 0$	$\neq 0$
Particle trajectories in a laboratory reference frame	A circle	A circle	Loop-like line

In each approximation, a non-uniform drift stream will be found from a given vorticity and the correction to the propagation velocity should be calculated already in the next approximation.

In a laboratory reference frame, the solution of the linear problem (6.11) has the form

$$X = a + \varepsilon u_1(b)t - \varepsilon \cdot \frac{1}{k}e^{kb}\sin k(a - c_0t);$$
$$Y = b + \varepsilon \cdot \frac{1}{k}e^{kb}\cos k(a - c_0t). \tag{6.13}$$

The wave is running to the right with velocity c_0. In the case $\sigma_1 = 0$, where there is no vorticity, Eq. (6.13) describes a linear potential wave (Stokes wave). For the Guyon wave, the function $\sigma_1(b) \neq 0$ (and is not constant); therefore, fluid particles, in addition to rotation around a circle, participate in a drift that is non-uniform in depth, so that their trajectory of motion is a loop-like line (see Table 6.1).

The quantity ε/k in Eqs. (6.11) and (6.13) means the linear wave amplitude A_1. The smallness of the parameter ε means a low steepness of the wave, $kA_1 \ll 1$. Note that in the case $\sigma_1 = 0$ and for a finite value of ε, Eqs. (6.11) and (6.13) are an exact solution of complete equations of fluid dynamics of an ideal fluid and describe a trochoidal wave (Gerstner wave, see Chapter 5). The following relations are valid for the Gerstner wave (see Eq. (5.7)):

$$\Omega_1 = 0; \quad \Omega_2 = 2kc_0e^{2kb}. \tag{6.14}$$

At low steepness, the Gerstner wave is vortex free and completely identical to the linear Stokes wave. In the limit $\varepsilon \ll 1$, the wave profile given by Eqs. (6.11) and (6.13) corresponds to a sine wave.

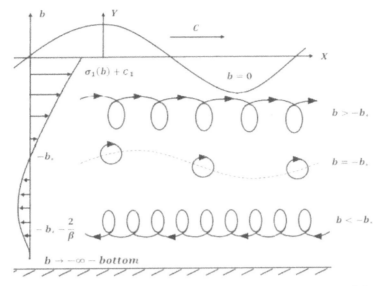

Fig. 6.2. Trajectories of fluid particles in a Guyon linear wave in a laboratory reference frame for the model $\sigma_1(b) + c_1 = \alpha(b + b_*)\exp\beta b$, $b_* > 0$.

The function $\sigma_1(b)$ in solution (6.11) describes the shear flow and, due to the arbitrariness of $\Omega_1(b)$, can be largely arbitrary too. In particular, the flow profile $\sigma_1(b)$ may have an inflection point. As is well known, linear waves on the flow surface $\sigma_0(b)$ with an inflection point on the profile are unstable [6]. But in the case considered here, $\sigma_0 = \text{const}$, and the flow profile $\sigma_1(b)$ is arbitrary. Let us assume

$$\sigma_1(b) = \alpha(b + b_*)\exp\beta b, \quad b_* > 0.$$

Here, the constant α has the dimension of velocity and the constant β has the dimension of the inverse length ($\alpha, \beta > 0$). The velocity of such a shear flow will be directed along the OX axis for particles with the coordinate $b > -b_*$ and against the axis in the case of inverse inequality. At point $b = -b_* - 2/\beta$, the profile has an inflection point. It is noteworthy that in a laboratory reference frame, particles located on the Lagrangian horizon $b = -b_*$ move in circles, and at higher and lower horizons along loop trajectories (see Fig. 6.2).

6.1.2. Quadratic approximation

We now write down the second-order equations of perturbation theory:

$$\xi_{2_q} + \eta_{2_b} = -\frac{D(\xi_1, \eta_1)}{D(q, b)}, \tag{6.15}$$

$$\sigma_0^2 \xi_{2_{qq}} + 2\sigma_0 \sigma_1 \xi_{1_{qq}} = -H_{2_q} + \frac{D(\eta_1, H_1)}{D(q, b)}, \tag{6.16}$$

$$\sigma_0^2 \eta_{2_{qq}} + 2\sigma_0 \sigma_1 \eta_{1_{qq}} = -H_{2_b} - \frac{D(\xi_1, H_1)}{D(q, b)}. \tag{6.17}$$

The unknown functions included in these equations should satisfy the boundary conditions

$$\sigma_0 \eta_2 + \sigma_1 \eta_1 \to 0 \quad \text{at } b \to -\infty, \tag{6.18}$$

$$\int_0^\lambda (\eta_2 + \eta_1 \xi_{1_q})\big|_{b=0} = 0, \tag{6.19}$$

$$H_2(q, 0) = g\eta_2(q, 0). \tag{6.20}$$

The quadratic terms in Eqs. (6.15)–(6.17) are easily calculated and are equal to

$$\frac{D(\xi_1, \eta_1)}{D(q, b)} = -e^{2kb}, \quad \frac{D(\eta_1, H_1)}{D(q, b)} = 0, \quad \text{and} \quad \frac{D(\xi_1, H_1)}{D(q, b)} = -kH_1^* e^{2kb}, \tag{6.21}$$

respectively. Let us differentiate Eq. (6.16) with respect to q, Eq. (6.17) with respect to the variable b, and add them up. Taking into account Eqs. (6.5), (6.15), and (6.21), we obtain

$$\Delta H_2 = 2k^2 \sigma_0^2 e^{2kb} + 2k\sigma_0 \sigma_1' e^{kb} \cos kq.$$

Solving this equation, we find

$$H_2 = \frac{1}{2}\sigma_0^2 e^{2kb} + H_{21}(b) \cos kq;$$

$$H_{21} = \sigma_0 \left(\sigma_1 e^{kb} - e^{-kb} \int_{-\infty}^b \sigma_1' e^{2kb} db \right). \tag{6.22}$$

According to the well-known function H_2, using Eqs. (6.16) and (6.17), we find the expressions for trajectory coordinates of fluid particles

$$\xi_2 = \frac{1}{k\sigma_0^2}(2\sigma_0\sigma_1 e^{kb} - H_{21})\sin kq; \qquad (6.23)$$

$$\eta_2 = \frac{1}{k\sigma_0^2}\left(\frac{1}{k}H'_{21} - 2\sigma_0\sigma_1 e^{kb}\right)\cos kq + \frac{1}{2k}e^{2kb}. \qquad (6.24)$$

These relations satisfy conditions (6.18) and (6.19). To fulfill the boundary condition (6.20) for pressure on a free surface, it is necessary that

$$\sigma_1(0) = \int_{-\infty}^{0} \sigma'_1 e^{2kb} db, \qquad (6.25)$$

or, with allowance for Eq. (6.12),

$$c_1 = -\int_{-\infty}^{0}(1 - e^{2kb})\Omega_1 db. \qquad (6.26)$$

It is noteworthy that on a free surface, the function H_{21} vanishes, as does the amplitude of the variable part of the elevation η_2 (cf. Eqs. (6.22) and (6.24)). The latter means that the wave profile does not change in the quadratic approximation, but the structure of the disturbances changes in the depth.

In Lagrangian coordinates, the solution of the second approximation is written as follows:

$$X = a - c_0 t + \varepsilon(\sigma_1 t + \xi_1) + \varepsilon^2(\sigma_2 t + \xi_2) + O(\varepsilon^3); \qquad (6.27)$$
$$Y = b + \varepsilon\eta_1 + \varepsilon^2\eta_2 + O(\varepsilon^3). \qquad (6.28)$$

The variable q in the expressions for ξ_1 and η_1 should be taken equal to $a - (c_0 - \varepsilon\sigma_1)t$, and in the expressions for ξ_2 and η_2 equal to $a - c_0 t$. In this approximation, a term containing the function $\sigma_2(b)$ is added to the shear flow. This function is related to the vorticity of the quadratic approximation

$$\sigma'_2 = -\Omega_2(b) + 2kc_0 e^{2kb}. \qquad (6.29)$$

Recall that the form of Ω_2 is assumed to be given. The function $\sigma_2(b)$ is determined up to a constant, the value of which, and hence

116   *Analytical Fluid Dynamics in Lagrangian Variables*

the value of the correction c_2 to the phase velocity of the wave, is calculated in the following approximation.

Equations (6.27) and (6.28) describe a nonlinear Guyon wave in the quadratic approximation with respect to the parameter ε. In essence, we solved the same problem that Guyon first set and studied [3], but we used a different method for this. Guyon employed the Eulerian approach and performed calculations in "coordinate x and a stream function ψ" variables. Our approach is based on calculations in modified Lagrangian variables, and it can be called "quasi-Lagrangian." Since vorticity and shear flows are functions of the vertical Lagrangian coordinate, rather than stream functions, as in Guyon's case, such a description looks more familiar and natural in terms of physics. In addition, in our case, it is convenient to compare the Guyon wave with the Stokes and Gerstner waves.

The difference in approaches leads to a difference in the form of representation of solutions of the corresponding approximations. Thus, Guyon's quadratic perturbation spectrum includes two harmonics [3,4]. It should also be noted that the functions $\sigma(b)$ and $\sigma(\psi)$ define different shear flow profiles. In this sense, not only the vertical perturbation structure but also the formulas for c_1 and higher-order phase velocity corrections have different representations for different approaches. At the same time, a certain choice of new Lagrangian coordinates (the vertical Lagrangian coordinate b is assumed to be equal to ψ) can combine the results of two approaches [7].

The Stokes wave is obtained from Eqs. (6.27) and (6.28) if one assumes $\Omega_1 = \Omega_2 = 0$ (that the function σ_1 is equal to zero follows from Eq. (6.25)). As a consequence, the Lagrangian representation of the Stokes wave will take the form

$$X = a - c_0 t - \varepsilon \cdot \frac{1}{k} e^{kb} \sin k(a - c_0 t) + \varepsilon^2 \sigma_2 t + O(\varepsilon^3); \quad (6.30)$$

$$Y = b + \varepsilon \cdot \frac{1}{k} e^{kb} \cos k(a - c_0 t) + \varepsilon^2 \cdot \frac{1}{2k} e^{2kb} + O(\varepsilon^3); \quad (6.31)$$

$$\sigma_2 = c_0 e^{2kb} - c_2. \quad (6.32)$$

The last of these relations is obtained by integrating Eq. (6.29) with allowance for condition (4.10). The first term on the right-hand side of Eq. (6.32) corresponds to the Stokes drift.

Condition (6.14) corresponds to a quadratically nonlinear Gerstner wave. By substituting the quadratic vorticity value in Eq. (6.29), we obtain $\sigma_2' = 0$. The velocity of the Gerstner wave coincides with the phase velocity c_0 of linear potential waves, so that all corrections to it are zero, $c_n = 0$, $n \geq 1$. Accordingly, the function σ_2 for the Gerstner wave is also zero, and its representation is written as follows:

$$X = a - c_0 t - \varepsilon \cdot \frac{1}{k} e^{kb} \sin k(a - c_0 t) + O(\varepsilon^3); \qquad (6.33)$$

$$Y = b + \varepsilon^2 \cdot \frac{1}{2k} e^{2kb} + \varepsilon \cdot \frac{1}{k} e^{kb} \cos k(a - c_0 t) + O(\varepsilon^3). \qquad (6.34)$$

These relations differ from the classical Gerstner solution (see Eq. (5.4)) by the presence of the second term in the expression for Y. The presence of this term in Eq. (6.34) is due to the features of the Lagrangian description. Lagrangian variables ("labels" of fluid particles) can be chosen in countless ways. It is possible, as is adopted in this section, to consider the coordinates a and b as the initial positions X_0 and Y_0 of the particles. Then, the right-hand side of the continuity equation in system (4.1) is equal to unity. However, such a requirement is not mandatory at all. It can be assumed that the initial coordinates of the particles are some functions of their labels, i.e., $X_0(a, b)$ and $Y_0(a, b)$. In this case, the continuity equation, unlike Eq. (4.1), will take the form

$$\frac{D(X, Y)}{D(a, b)} = \frac{D(X_0, Y_0)}{D(a, b)} \neq 1. \qquad (6.35)$$

The only constraint on the form of the right-hand side of Eq. (6.35) is that the sign of the Jacobian should be constant in the flow region; more precisely, the Jacobian should not vanish. Otherwise, the unambiguity of the flow representation is violated.

For the classical Gerstner solution, $X_0 \neq a$ and $Y_0 \neq b$, and the continuity equation is written as Eq. (6.35). As a consequence of this difference, an additional term appears in Eq. (6.34) in our description. At the same time, the conventional form of the Gerstner solution can be achieved by making the replacement

$$b \to b + \varepsilon^2 \cdot \frac{1}{2k} e^{2kb}$$

in Eqs. (6.33) and (6.34).

6.1.3. Cubic approximation

The complete solution in the cubic approximation is given in a different study [5]. As the most important result, we indicate the formula for the quadratic addition to the wave propagation velocity:

$$c_2 = \frac{1}{2}c_0 - \int_{-\infty}^{0} \Omega_2(1 - e^{2kb})db. \qquad (6.36)$$

It is noteworthy that it does not depend on the first approximation vorticity Ω_1. For the Stokes wave, $\Omega_2 = 0$, and the quantity c_2 is half the c_0 value, as should be expected. For the Gerstner wave, $c_2 = 0$, also in full accordance with the theory. For Guyon waves, the quantity c_2 can take a variety of values, depending on the distribution of the vorticity Ω_2.

The Guyon wave propagates with velocity

$$c = c_0 + \varepsilon c_1 + \varepsilon^2 c_2 + O(\varepsilon^3), \qquad (6.37)$$

where linear velocity corrections are given by Eqs. (6.26) and (6.36).

To develop this line of research, the effect of the finite depth of the fluid on Guyon waves was examined [8]. Solutions for the first two approximations are found. Thus, Guyon's result was generalized to the case of finite depth.

6.2. Development of the Modified Coordinates Method

Modified Lagrangian coordinates can also be used to study more complex motions of waves with a free surface.

6.2.1. Stationary spatial motion

Let us ask ourselves the following: "Is it possible to generalize two-dimensional Guyon waves to the three-dimensional case?" Representation (6.2) will obviously be invalid. Vorticity is not preserved in space flows. Meanwhile, modified Lagrangian variables can again be employed to analyze three-dimensional vortex waves [9].

Consider three-dimensional stationary waves in an infinitely deep fluid. Let a wave with a constant profile propagate along the x-axis. Let us switch to a reference frame moving at the speed of the wave.

In this reference frame, the motion will be stationary and the wave profile will be fixed.

We introduce spatial modified Lagrangian variables $q = a + \sigma(b)t, b, c$ and use them for writing three-dimensional equations of fluid dynamics (1.11), (1.18):

$$\frac{D(X, Y, Z)}{D(q, b, c)} = \frac{D(X_0, Y_0, Z_0)}{D(q, b, c)} = D_0(b, c), \tag{6.38}$$

$$\sigma^2(X_{qq}X_q + Y_{qq}Y_q + Z_{qq}Z_q) = -\frac{1}{\rho}p_q - gY_q, \tag{6.39}$$

$$\sigma^2(X_{qq}X_b + Y_{qq}Y_b + Z_{qq}Z_b) = -\frac{1}{\rho}p_b - gY_b, \tag{6.40}$$

$$\sigma^2(X_{qq}X_c + Y_{qq}Y_c + Z_{qq}Z_c) = -\frac{1}{\rho}p_c - gY_c. \tag{6.41}$$

By integrating Eq. (6.39), we obtain the Bernoulli integral

$$\frac{\sigma^2}{2}(X_q^2 + Y_q^2 + Z_q^2) + gY + \frac{p}{\rho} = B(b, c). \tag{6.42}$$

Equations (6.39)–(6.41) have three more Cauchy invariant integrals (see Section 1.3.3):

$$\frac{D(\sigma X_q, X)}{D(b, c)} + \frac{D(\sigma Y_q, Y)}{D(b, c)} + \frac{D(\sigma Z_q, Z)}{D(b, c)} = S_1(b, c); \tag{6.43}$$

$$\sigma\left[\frac{D(X_q, X)}{D(c, q)} + \frac{D(Y_q, Y)}{D(c, q)} + \frac{D(Z_q, Z)}{D(c, q)}\right] = S_2(b, c); \tag{6.44}$$

$$\sigma\left[\frac{D(X_q, X)}{D(q, b)} + \frac{D(Y_q, Y)}{D(q, b)} + \frac{D(Z_q, Z)}{D(q, b)}\right]$$

$$-\sigma'(X_q^2 + Y_q^2 + Z_q^2) = S_3(b, c). \tag{6.45}$$

All written equations and relations do not explicitly include time. Systems (6.38)–(6.45) can be solved by the perturbation theory method. In another study [9], Stokes expansions for the first two approximations were constructed for the problem of Guyon spatial waves in deep water. The found approximate solution is a non-trivial example of spatial vortex flow with complex trajectories of fluid particles.

6.2.2. Non-stationary flow

For simplicity, we focus on two-dimensional non-stationary motion. In the coordinates q, b, and t, the system of Eq. (4.1) will take the form

$$X_q Y_b - X_b Y_q = 1,$$

$$\sigma^2 X_{qq} + 2\sigma X_{qt} + X_{tt} = \frac{D(Y, P)}{D(q, b)},$$

$$\sigma^2 Y_{qq} + 2\sigma Y_{qt} + Y_{tt} = \frac{D(P, Y)}{D(q, b)}.$$

This system of equations was used, in particular, to describe vortex standing waves in deep water [10].

The perturbations ξ and η were sought in the form of series with respect to a small parameter of wave steepness (by analogy with Eqs. (4.13), (4.14), and (6.4)). Obviously, secular terms will not arise in such a calculation procedure.

A weak (of order ε) shear flow $\varepsilon\sigma_1(b)$ is imposed on periodic fluctuations of the fluid. If the condition $\sigma_1(0) = 0$ is imposed ($b = 0$ corresponds to a free surface), then the waves on the surface will be observed as standing ones. At the same time, wave nodes drift slowly in the depth. In yet another study [10], calculations for the first three approximations were performed. It is interesting that in the third order of perturbation theory, vorticity, in addition to the "drift" component $-\sigma_3'$, also contains a term depending on σ_1. The presence of this component of vorticity, unlike the first two approximations, serves as a manifestation of the effect of the nonlinear interaction of the wave with the shear flow.

6.3. Nonlinear Schrödinger Equation for Quadratic Vortex Waves

Among the class of Guyon waves, the family of quadratic vortex waves, for which $\Omega_1 = 0$ and $\Omega_2(b) \neq 0$, deserves special attention. Despite their more specific form, they generalize both the Stokes and the Gerstner waves. The dispersion relation for these waves is written as follows (see Eqs. (6.36) and (6.37)):

$$\omega = \sqrt{gk}(1 + \gamma k^2 A^2); \tag{6.46}$$

$$\gamma = \frac{1}{2} - \int_{-\infty}^{0} \Omega_2 c_0^{-1}(1 - e^{2kb})db. \tag{6.47}$$

If $\Omega_2 = 0$, then $\gamma = 1/2$, and Eq. (6.46) corresponds to the Stokes wave (see Eq. (4.18)). If $\Omega_2 = 2kc_0e^{2kb}$ (see Eq. (5.7)), then $\gamma = 0$, and Eq. (6.46) coincides with the dispersion relation of Gerstner waves. For Guyon waves, the γ values can be arbitrary, both positive and negative. According to the Lighthill criterion, weakly nonlinear wave trains of Guyon waves will be unstable if $\omega_0'' \gamma < 0$ [11, 12]. Since $\omega_0'' = -\omega_0/4k^2$, the instability condition will be $\gamma > 0$. Stokes waves are always unstable. Gerstner waves do not satisfy this criterion, and it is impossible to talk about their instability in this (cubic) approximation.

Based on the nonlinear dispersion relation (6.46), following the procedure described in Section 4.2, it is possible to write the equation of evolution of the envelope of the Guyon wave train [5]:

$$i\left(A_t^* + \frac{\omega_0}{2k_0}A_q^*\right) - \frac{\omega_0}{8k_0^2}A_{qq}^* - \gamma\omega_0 k_0^2|A^*|^2 A^* = 0, \tag{6.48}$$

where $A^* = A\exp i\theta$ is the complex amplitude of the train envelope, $A = |A^*|$ is the wave amplitude, and θ is the phase. The variable q plays the role of X, and these can be replaced by each other when writing the nonlinear Schrödinger Eq. (6.48). Equation (6.48) is also obtained by the multiscale expansion method [1].

Let us consider in more detail the problem of the stability of a homogeneous wave train within the framework of Eq. (6.48). We introduce dimensionless variables $t_* = \omega_0 t$ and $q_* = k_0(q - c_g t)$ (transition to a system moving with group velocity); then, Eq. (6.48) will take the form

$$iA_{t_*}^* - \frac{1}{8}A_{q_*q_*}^* - \gamma k_0^2|A^*|^2 A^* = 0. \tag{6.49}$$

This equation is satisfied by expressions describing a homogeneous wave train, namely, $A = A_0 = \text{const}$ and $\theta_0 = -\gamma k_0^2 a^2 t_*$. We introduce perturbations of the real quantities $\tilde{A}$ and $\tilde{\theta}_x$ using the formulas

$$A = A_0 + \tilde{A}, \quad \theta_x = \theta_{0x} + \tilde{\theta}_x = \tilde{\theta}_x.$$

Then, the system of linearized Eq. (6.50) can be reduced to an equation for $\tilde{A}$ of the following form:

$$\tilde{A}_{t_* t_*} + \frac{1}{8}\left(\frac{1}{8}\tilde{A}_{q_* q_* q_* q_*} + 2\gamma k_0^2 A_0^2 \tilde{A}_{q_* q_*}\right) = 0.$$

By substituting perturbations $\tilde{A} \sim \exp i(\Omega^* t_* + K q_*)$, we find that the perturbation instability condition $\Omega^{*2} < 0$ has the form

$$0 < K < 4k_0 A_0 \sqrt{\gamma}. \tag{6.50}$$

The maximum increment corresponds to the value $K = 2k_0 A_0 \sqrt{2\gamma}$. Inequality (6.50) was obtained in another way in a different study [13].

As for the train of potential waves (see Section 4.2), the coefficient of the nonlinear term in the NLSE is related to the drift of particles. Depending on the sign of γ, Eq. (6.49) can be either focusing ($\gamma > 0$ and there is a modulation instability) or defocusing (the case is the opposite). The boundary case $\gamma = 0$ corresponds to the Gerstner wave, which does not satisfy the Lighthill criterion. In general case equation (6.48) can be used to study rogue waves [14].

References

[1] Abrashkin, A.A. and Pelinovsky, E. (2017). Lagrange form of the nonlinear Schrödinger equation for low-vorticity waves in deep water, *Nonl. Proc. Geophys.*, 24, pp. 255–264.

[2] Abrashkin, A.A. and Pelinovsky, E.N. (2018). Dynamics of a wave packet on the surface of a non-uniformly vortex fluid (Lagrangian description). *Izvestiya, Atmospheric and. Oceanic Physics*, 54(1), pp. 101–105.

[3] Gouyon, R. (1958). Contribution à la Théorie des Houles. *Annales de la Faculté des Sciences de l'Université de Toulouse*, 4, 22, pp. 1–55.

[4] Sretensky, L.N. (1977). *Theory of the Fluid Wave Motions* [in Russian], (Nauka, Moscow).

[5] Abrashkin, A.A. and Pelinovsky, E.N. (2021). Nonlinear Goyuon waves. *J Phys. A: Math. Theory.*, 395701, pp. 19.

[6] Drazin, P.G. (2002). *Introduction to Hydrodynamic Stability* (Cambridge University Press, Cambridge).

[7] Clamond, D. (2007). On the Lagrangian description of steady surface gravity waves. *J. Fluid Mech.*, 589, pp. 434–454.

[8] Abrashkin, A.A. (2022). Goyuon waves in water of finite depth. *Monatshefte für Mathematik*, 199, 4, pp. 717–732.

[9] Abrashkin, A.A. (1996). Three-dimensional Gouyon waves. *Fluid Dynamics*, 31(4), pp. 583–587.

[10] Abrashkin, A.A. (1996). Standing vortex waves in deep water. *Fluid Dynamics*, 31(3), pp. 470–473.

[11] Lighthill, M.J. (1965). Contribution to the theory of waves in nonlinear dispersive systems. *Journ. Inst. Math. Appl.*, 1, pp. 269–306.

[12] Whitham, G.B. (1974). *Linear and Nonlinear Waves* (A Wiley-Interscience Publication John Wiley & Sons, New York).

[13] Pizzo, N. Lenain, L. Luc, Rømcke, O., Ellingsen, S.Å. and Smeltzer, B.K. (2023). The role of Lagrangian drift in the geometry, kinematics and dynamics of surface waves. *J. Fluid Mech.*, 954, R4.

[14] Kharif, C., Pelinovsky, E. and Slunyaev, A. (2009). *Rogue Waves in the Ocean* (Springer-Verlag, Berlin).

Part III
Exact Solutions in Vortex Dynamics

Chapter 7

Two-Dimensional Vortex Flow

7.1. Review of Known Approaches in the Study of Plane Vortex Flows

> All exact science is based on the
> idea of approximation.
>
> — Bertrand Russell

Two-dimensional equations of the dynamics of an ideal incompressible fluid in the "vorticity - stream function" variables have the form [1]

$$\frac{d\Omega}{dt} = 0; \quad \Delta\psi = -\Omega. \tag{7.1}$$

From these equations, it follows, in particular, that if the vorticity is constant everywhere, then the problem is reduced to solving the Poisson equation, for which it is easy to write down the general solution. Thus, the problem of describing the uniformly vortical motion of a fluid is related to the choice of a solution satisfying specified boundary conditions. So, in the case of fluid motion inside a solid body, this problem is solved for an elliptical cylinder [2] (see Section 3.3.3) and a triangular prism [3]. In the case of a flow limited by a plane, both known exact solutions were obtained by Chaplygin. Others describe the case where a flow that is uniform at infinity moves around an obstacle in the form of a round semi-cylinder [4] and the form of a fence [5].

Another type of motion with constant vorticity is localized vortex regions located in the surrounding potential flow. These regions include the Rankin circular vortex [1, 6] and the Kirchhoff elliptical vortex [7] located in a potential flow at rest at infinity (see Chapter 3), the Chaplygin elliptical vortex located in a shear flow [8], and also elliptical vortices, namely, the stationary Saffman–Moore vortex [9, 10] and the non-stationary Kida vortex [11] located in an external field linear to coordinates. A specific feature of these flows is the presence of a boundary separating the areas of the vortex and the potential flows. The problems of their description will be discussed in Chapter 9.

In stationary motion, the trajectories of fluid particles coincide with the streamlines. Since the vorticity of particles in a plane flow is preserved, it will be constant along the streamlines. It follows that Eq. (7.1) is reduced to the following relation:

$$\psi_{xx} + \psi_{yy} = -\Omega(\psi), \tag{7.2}$$

where $\Omega(\psi)$ is some function that depends only on ψ. Thus, if the distribution of vorticity along the streamlines is known, the mathematical formulation of the problem of steady two-dimensional flow is clear, although solving Eq. (7.2) may not be easy.

Vorticity distribution can be specified as arbitrary if we limit ourselves to the theory of inviscid fluid. In practice, the distribution is determined by the history of the formation of a steady flow, which, in turn, largely depends on viscosity. It is usually not possible to analyze in detail the process of the formation of steady motion; therefore, the function $\Omega(\psi)$ can be determined only in some simple cases. In this sense, the epigraph to the chapter can be interpreted as follows: Viscosity is involved in the formation of a stationary vortex flow, but we will use the approximation of an ideal fluid and search for exact solutions.

Possible solutions to the momentum equations for an inviscid fluid can be studied by specifying the shape of $\Omega(\psi)$. Thus, in problems of flow around solid bodies with the formation of an attached vortex, vorticity can be considered constant. The linear function $\Omega(\psi)$ corresponds to a vortex flow inside a circlular Chaplygin dipole vortex located in an external uniform flow [12]. The boundary separating the equally swirling parts of a circular core can be either the diameter of the circle (symmetric dipole vortex) or an arc based on the diameter (asymmetric dipole vortex). The first of these was also described independently by Lamb [1] and is often mentioned in the literature by his name. Everywhere, the regular solution (7.2), corresponding to the exponential function $\Omega(\psi)$, describes the path of circular vortices in a shear flow [13]. This flow is vortical everywhere, and the boundaries of vortices are defined as lines within which the vorticity exceeds a certain specified value.

In a study [14], Shercliff considered flows for which the dependence of vorticity on the stream function is given by the formula

$$\Omega(\psi) = \alpha\psi \ln \psi + \beta\psi, \tag{7.3}$$

where α and β are constants. Solutions (4.2) for the right-hand side of Eq. (7.3) are found by the method of separation of variables. The corresponding flows can be of two types, namely, "bending" (not containing closed streamlines) and "loop" (containing closed streamlines). Their "geometry" can be very different. The flows bounded by rectilinear boundaries are most interesting among them. Equation (7.3) is the only one that permits the separation of variables in the form of a product of the functions x and y.

In the development of this study [14], Kaptsov found all the representations of the right-hand sides of Eq. (4.2), for which the solution can be represented as

$$\psi = Q[\xi(x) + \eta(y)],$$

where Q, ξ, η are some functions, and none of them is a polynomial of degree two or less [15, 16].

Cherny studied a class of flat stationary self-similar flows in detail, the vorticity of which depends strongly on ψ [17].

Among the non-stationary vortex flows described by the exact solution (4.1), we mention an asymmetric Chaplygin dipole vortex moving along a circular trajectory in a fluid at rest at infinity [12]. The Chaplygin solution was rediscovered in a study [18], where the formation of a dipole structure by a jet of uniformly rotating fluid was analyzed.

This brief review concerns the Eulerian description of a fluid. A list of well-known analytical descriptions of vortices in Lagrangian variables is given in Chapter 3. We now try to extend it.

7.2. Ptolemaic Fluid Motion Class

By direct substitution in Eqs. (3.11) and (3.12), it can be verified that the expression

$$W = G(\chi)\exp(i\lambda t) + F(\bar{\chi})\exp(i\mu t), \qquad (7.4)$$

where G and F are analytical functions, and λ and μ are real numbers, is the exact solution of the equations of two-dimensional fluid dynamics [19, 20]. The functions G and F are largely arbitrary since the only restriction on their choice is the non-vanishing of the Jacobian D (see Chapter 1), which has the form

$$D = |G'_\chi|^2 - |F'_{\bar{\chi}}|^2; \quad G'_\chi = \frac{dG}{d\chi}, \quad F'_{\bar{\chi}} = \frac{dF}{d\bar{\chi}}. \qquad (7.5)$$

The trajectories of fluid particles for motion (7.5) are epicycloids (hypocycloids), i.e., the particles describe a circle, the center of which moves along another circle (see Fig. 7.1). Planets in the Ptolemaic worldview revolved along such orbits. In this regard, this type of flow was called Ptolemaic. An arbitrary time function describing the fluid motion as a whole can be added to Eq. (7.4). But for the sake of simplicity, we will not do this.

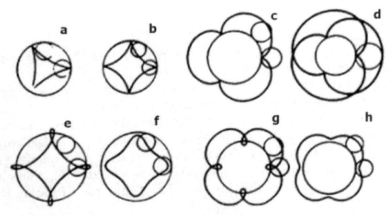

Fig. 7.1. Possible types of trajectories of fluid particles in Ptolemaic flows: (a) and (b) hypocycloids; (c) and (d) epicycloids; (e) an elongated hypocycloid with four protrusions ($\mu/\lambda = -3$); (f) a shortened hypocycloid; (g) an elongated epicycloid; and (h) a shortened epicycloid ($\mu/\lambda = 4$).

Two-Dimensional Vortex Flow

Ptolemaic solutions (7.4) describe a very wide class of two-dimensional vortex motions. Consider some of their common properties. It is obvious that the uniform rotation of the fluid as a whole with angular velocity ω_* is characterized by a common factor $\exp(i\omega_* t)$ in the expression for W. Therefore, by choosing an appropriate reference frame, the factors for the functions G and F can be changed. With this procedure, however, the frequency difference $\lambda - \mu$ will remain unchanged. In particular, one of the terms in Eq. (7.4) can be made time-independent.

Ptolemaic flows are vortical. Using Eq. (3.12), we find the following vorticity vector for them:

$$\vec{\Omega} = 2\frac{\lambda|G'_\chi|^2 - \mu|F'_{\bar{\chi}}|^2}{|G'_\chi|^2 - |F'_{\bar{\chi}}|^2}\vec{z}_0. \tag{7.6}$$

The vorticity vector is normal to the plane of motion of fluid particles and, in general, has a non-uniform distribution.

Ptolemaic flows include two motions that we have already considered. These are the Kirchhoff vortex and the Gerstner waves. For the first of them, the functions G and F are linear, and the frequency in the second term is zero (see Eq. (3.21)). The complex representation for Gerstner waves is obtained from Eq. (5.4) and has the following form:

$$W = \chi + iA \exp i(k\bar{\chi} - \omega t). \tag{7.7}$$

Here, one of the time frequencies is missing again, but a function of F is already the exponent.

Let us find the Eulerian velocity field corresponding to the Ptolemaic flow. We now write down an expression for the complex flow velocity:

$$V = i\lambda G(\chi)\exp(i\lambda t) + i\mu F(\bar{\chi})\exp(i\mu t). \tag{7.8}$$

The next step will be to exclude the variables χ and $\bar{\chi}$ from the system of Eqs. (7.4) and (7.8) and determine the relation of the complex velocity V with complex coordinates W and $\bar{W}$. This cannot be done in the general case, but let the relation $F = Q(\bar{G})$ be known. Then, solving Eqs. (7.4) and (7.8) with respect to the functions

132   *Analytical Fluid Dynamics in Lagrangian Variables*

G and F, we obtain

$$G = \frac{i\mu W - V}{i(\mu - \lambda)} e^{-i\lambda t}, \quad F = \frac{V - i\lambda W}{i(\mu - \lambda)} e^{-i\mu t}.$$

As a result, the velocity as a function of coordinates will be implicitly determined by the formula

$$\frac{V - i\lambda W}{i(\mu - \lambda)} e^{-i\mu t} = Q \left[\frac{i\mu \bar{W} + \bar{V}}{i(\mu - \lambda)} e^{i\lambda t} \right].$$

This function can be rewritten in the following symmetric form:

$$F^{-1} \left(\frac{V - i\lambda W}{i(\mu - \lambda)} e^{-i\mu t} \right) = G^{-1} \left(\frac{i\mu \bar{W} + \bar{V}}{i(\mu - \lambda)} e^{i\lambda t} \right),$$

where F^{-1} and G^{-1} are the inverse functions of G and F, respectively. It is obvious that, despite the simplicity of the time dependence on particle trajectories, the flow field in Eulerian variables can be a complex function.

The pressure in the Ptolemaic flow region is written as follows:

$$\begin{aligned}
\frac{p}{\rho} &= -H + \frac{1}{2}\lambda^2 |G|^2 + \frac{1}{2}\mu |F|^2 \\
&\quad + \mathrm{Re} \left\{ e^{i(\lambda - \mu)t} \int [\lambda^2 G \bar{F}'_\chi + \mu^2 G'_\chi \bar{F}] \right\} d\chi,
\end{aligned} \tag{7.9}$$

where H is the potential of external forces.

7.3. The Uniqueness of Ptolemaic Flows

In regard to the type of solution (7.4), a natural question arises: "Is it possible to generalize the Ptolemaic solutions by adding epicycles, just as Ptolemy did to describe the motion of the planets" In other words, is it possible to have solutions to the equations of fluid dynamics in the form of a finite series, i.e., when the complex coordinate is represented as

$$W = \sum_{k=1}^{N} W_k(\chi, \bar{\chi}) \exp(i\lambda_k t), \tag{7.10}$$

where the number N of terms is finite (more than two), and all λ_k are real?

We show that of all possible solutions in Lagrangian variables containing a finite set of time (coordinate-independent) frequencies, only a two-frequency solution describing Ptolemaic flows satisfies the equations of fluid dynamics [21]. Let us consider successively the possibility of the existence of a solution for $N = 2, 3$, and, finally, for an arbitrary finite number.

So, let $N = 2$. In this case, the right-hand sides of the system of Eqs. (3.11) and (3.12) are independent of time only if

$$W_{1\chi}\overline{W_{2\bar\chi}} = W_{1\bar\chi}\overline{W_{2\chi}}.$$

This equality is equivalent to the following:

$$\frac{D(W_1, \overline{W_2})}{D(\chi, \bar\chi)} = [W_1, \overline{W_2}] = 0.$$

In the future, we will use square brackets for obtaining the Jacobian. It follows from the equality of the Jacobian to zero that

$$W_1 = f(\overline{W_2}), \tag{7.11}$$

where f is an analytical function. Before moving on to the case $N = 3$, we make a small digression. Equation (7.11) suggests that a more general type of exact solution of the fluid dynamics equations than Eq. (7.4) should be chosen. Mathematically, this is an important point. However, as will be shown, to solve specific problems, one can start with Eq. (7.4) without loss of generality and assume $W_1 = G(\chi)$. This is due to the possibility of arbitrary (unique) labeling of fluid particles and free choice of the variation range of Lagrangian coordinates, or, even deeper, due to relabeling symmetry of the Lagrangian description.

It also follows from the above-mentioned analysis that in the case of a finite sum (7.10), the terms containing the maximum and minimum exponents (let them be λ_1 and λ_N) are connected by the relation

$$W_1 = f(\overline{W_N}). \tag{7.12}$$

We now consider $N = 3$, where

$$W = \sum_{k=1}^{3} W_k \exp(i\lambda_k t). \tag{7.13}$$

134   *Analytical Fluid Dynamics in Lagrangian Variables*

Let us substitute this expression into the continuity equation and equate all time-oscillating terms to zero. There are two possible cases here:

(a) All frequency differences are not equal to each other. Then, we obtain

$$[W_1, \overline{W_2}] = 0, \quad [W_1, \overline{W_3}] = 0, \quad \text{and} \quad [W_2, \overline{W_3}] = 0. \quad (7.14)$$

It follows from the first condition that $\overline{W_2}$ is a function of W_1, from the second that $\overline{W_3}$ is a function of W_1, but then from the last and first equalities, we obtain that $\overline{W_1}$ is a function of W_1. This is possible if W_1 is a complex function of one real parameter. Obviously, W_2 and W_3 are also functions of the same parameter. Let us take it for the Lagrangian variable a. Finally, we come to the statement that W is a function of only a and thereby prove that expression (7.13) with a non-equidistant spectrum cannot be a solution to two-dimensional equations of fluid dynamics.

(b) However, let the intervals in the frequency triplet be the same (all λ_k are different):

$$\lambda_2 - \lambda_1 = \lambda_3 - \lambda_2.$$

Now, it is necessary to use both equations of fluid dynamics. It follows from the continuity equation that

$$[W_1, \overline{W_3}] = 0 \quad \text{and} \quad [W_1, \overline{W_2}] + [W_2, \overline{W_3}] = 0,$$

and from the momentum equation that

$$\lambda_1 [W_1, \overline{W_2}] + \lambda_2 [W_2, \overline{W_3}] = 0.$$

Here again, as in case (a), we obtain a set of conditions (7.14), and again W is a function of only a.

Finally, we turn to the case of an arbitrary finite number of terms in the sum (7.10). One of the relations between the functions W_k is the condition (7.12). We now arrange in the non-increasing order all the frequency differences, from the maximum $\lambda_N - \lambda_1$ to the difference that is closest to the frequency $(\lambda_N - \lambda_1)/2$, but is greater than it. Jacobians composed of pairs of functions corresponding to pairs of frequencies included in these differences must be zero. This follows from the continuity equation if there are no equal frequency differences, or from the system of equations of fluid dynamics if there are pairs of equal frequencies (see the triplet argument). It follows that functions W_k with frequencies, located on the frequency axis on one side of the value $(\lambda_N - \lambda_1)/2$, will functionally depend on complex conjugate functions $\overline{W_j}$, which correspond to frequencies from the other half of the frequency range. It is easy to see that in this case, any pair of functions from one-half of the frequency range is functionally interconnected (without complex conjugation).

Let us choose the function W_p with frequency λ_p closest to the value $(\lambda_N - \lambda_1)/2$. For the sake of certainty, let it be located in the same half of the interval as W_1. Then, according to the last statement,

$$W_1 = f_1(W_p),$$

as the frequencies are located in one-half of the frequency range. However, from the continuity equation (see arguments about the doublet and triplet), we have

$$W_1 = f_2(\overline{W_p}).$$

The last two equalities mean that W_1 is a function of only one real parameter a. Obviously, all other W_p are also just functions of this parameter. Let us take it as a Lagrangian coordinate. As a result, we conclude that W is only a function of a, which means that Eq. (7.10) for $N > 3$ cannot be a solution to two-dimensional equations of fluid dynamics of an ideal incompressible fluid. Thus, we have shown that of all possible solutions in Lagrangian variables containing a finite set of time (coordinate-independent) frequencies, only a two-frequency solution describing Ptolemaic flows satisfies the equations of fluid dynamics.

The solution of the form (7.10) cannot be generalized to the case of three or more frequencies. The requirements for the finiteness and discreteness of the frequency spectrum of a plane flow turn out to be quite stringent. General results of this type in relation to the Euler equation are unknown to us. According to the scenario of transition to the Landau–Hopf turbulence after the successive development of a number of unstable regimes, the spectrum of a fluid dynamic flow can be represented as a finite set of incommensurable frequencies. The superposition of the oscillations determined by these frequencies will, generally speaking, give a complex non-repeating picture of motion, which is proposed to be identified with turbulence. This kind of flow is described by Eq. (7.10) if we assume that the ratio of at least one pair of frequencies λ_k is an irrational number. The feature of such a flow is that fluid particles move in a bounded area and are not carried away by the flow. It is believed that, due to the oscillation synchronization phenomenon, the probability of the Landau–Hopf scenario is extremely low [22]. Our study proves that this scenario is fundamentally impossible for two-dimensional plane flows with localized trajectories of fluid particles.

Equation (7.10) defines flows with periodic or quasi-periodic (in the case of incommensurable frequencies) trajectories of fluid particles. This equation lacks a term describing the average (shear) flow. However, it does not include the circular rotation of the fluid, which is described by the expression $W = \chi \exp[i\omega_*(|\chi|^2)t]$, where $\omega_*(|\chi|^2)$ is the angular velocity of rotation of fluid particles. These observations open up two more possible directions for generalizing Ptolemaic flows, namely, the inclusion of a linear term in the formula for the complex coordinate of the trajectory W and the introduction of the dependence of frequencies λ and μ on Lagrangian coordinates in Eq. (7.10). But both directions, as we found out, do not give results.

Other researchers have also studied the properties of Ptolemaic flows. One book [16] provides a group analysis of solutions (7.4). Other papers [23,24] analyze the properties of harmonic maps underlying exact solutions of the form

$$X(t; a, b) + iY(t; a, b) = G_*(t, \chi) + \overline{F_*(t, \chi)}, \quad \chi = a + ib.$$

They include Ptolemaic flows, as well as a class of exact solutions with linear spatial coordinate functions G_* and F_* [20], which are discussed in Chapter 9. The original representation of Ptolemaic flows in real form (using matrices) is given in other studies [25,26]. At the same time, the authors of the study [25] wrongfully called new the solutions of (7.10) and (7.11) for $N = 2$.

7.4. Gerstner Waves on a Cylindrical Surface

Consider wave motion on the surface of a fluid partially filling a cylinder (centrifuge) rotating rapidly with frequency ω around the horizontal axis. Under the action of centrifugal force, the fluid is

Two-Dimensional Vortex Flow

pressed against the wall of the cylinder and moves with it around the central air core. The waves that arise in this case are called centrifuge waves [27] and were previously studied only in a linear approximation [27,28]. However, if the outer radius of the centrifuge is considered infinite (sufficiently large or much larger than the size of the air core R), the problem can be solved exactly [29].

Assume that the waves under study belong to the class of Ptolemaic flows. Then, the functions G and F should be determined by boundary conditions. Since the fluid rotates as a whole at infinity, in solution (7.4), we should assume

$$G(\chi) = R \exp\left(ik\chi\right) = \nu, \quad \lambda = \omega, \tag{7.15}$$

and consider that $|F| \to 0$ for $|\nu| \to \infty$.

We find the function F from the condition of constant pressure on a free surface, $|\nu| = R$. With allowance for Eq. (7.15), the expression for pressure (7.9) can be represented as follows:

$$\frac{p}{\rho} = \frac{\omega^2}{2}|\nu|^2 + \frac{\mu^2}{2}|F|^2 + \mathrm{Re}\int(\omega^2\nu\bar{F}' + \mu^2\bar{F})d\nu e^{i(\omega-\mu)t}.$$

To ensure that pressure on a free surface remains constant and is equal to p_0, it is necessary that the time multiplier coefficient vanishes. This is fulfilled if

$$F(\bar{\nu}) = A\bar{\nu}^{-q^2}, \quad q = \frac{\mu}{\omega}, \tag{7.16}$$

where A is the constant. We obtain the final expression for W by substituting (7.15) and (7.16) into (7.4):

$$W = \nu\exp\left(i\omega t\right) + A\bar{\nu}^{-q^2}\exp(iq\omega t), \tag{7.17}$$

from where it is easy to conclude that the trajectories of fluid particles are shortened epi- ($q > 0$) and hypocycloids with $|q - 1|$ protrusions, while the profiles of propagating waves are epicycloids with $q^2 - 1$ cusps. Herein, for the stationary profile, q^2 should be considered an integer. With allowance for the type of profile, it is natural to call the waves epicycloidal or, considering the connection between trochoids and epicycloids, Gerstner waves on a cylindrical surface.

The pressure distribution in a fluid is given by the formula

$$p = p_0 + \frac{\rho\omega^2}{2}[q^2A^2(|\nu|^{-2q^2} - R^{-2q^2}) + |\nu|^2 - R^2]. \tag{7.18}$$

For the constant A, which determines the amplitude of the waves, there is an upper limit R^{q^2+1}/q^2 when the profile of the free surface contains cusps (at high values of A, loops form on the profile, and this case is physically impossible).

Epicycloidal waves are vortex waves. The vorticity for them is written as follows (see Eq. (3.12)):

$$\Omega = \frac{2\omega(1 - q^5A^2|\nu|^{-2(q^2+1)})}{1 - q^4A^2|\nu|^{-2(q^2+1)}},$$

from where it can be seen that the vorticity depends, among other things, on the sign of q.

Let us find the angular rotation velocity ω_0 of the wave profile. Obviously, the rotation of the fluid as a whole with frequency ω_0 is characterized by a common factor $\exp(i\omega_0 t)$ in the expression for W; therefore, in the reference frame where the profile is stationary, the solution (7.17) will take the form

$$W = \nu \exp[i(\omega - \omega_0)t] + A\bar{\nu}^{-q^2}\exp[i(q\omega - \omega_0)t].$$

In this system, the trajectories of fluid particles coincide with the shape of the profile, so the equality

$$q\omega - \omega_0 = q^2(\omega - \omega_0)$$

is valid, from which we find

$$\omega_0 = \frac{q\omega}{q+1}.$$

In a reference frame moving with angular velocity ω, the rotation frequency of the profile is $(q+1)^{-1}\omega$, so that the waves corresponding to negative q move in the direction of rotation and those corresponding to positive q move in the opposite direction.

It is interesting to note that the solution (7.17) will remain valid in a fluid with an arbitrary stratification $\rho_*(|\nu|)$. Indeed, we assume that

$$
\begin{aligned}
\frac{1}{\rho_*(|\nu|)}\frac{\partial p_*}{\partial a} &= \frac{1}{\rho}\frac{\partial p}{\partial a}; \\
\frac{1}{\rho_*(|\nu|)}\frac{\partial p_*}{\partial b} &= \frac{1}{\rho}\frac{\partial p}{\partial b},
\end{aligned}
\tag{7.19}
$$

where p is given by (7.18). In this case, all the arguments for a homogeneous fluid are automatically transferred to an inhomogeneous one. Since p is only a function of $|\nu|$, the representation of (7.19) is always possible.

The expression for pressure in a stratified fluid is written as

$$
p_* = \int \rho_*(|\nu|)\omega^2(|\nu| - A^2|\nu|^{-2q^2-1})d|\nu|.
\tag{7.20}
$$

Thus, the presence of stratification does not change the shape of particle trajectories in epicycloidal waves, as well as the frequencies of the motion. In addition, isopycnic surfaces ($\rho_* = $ const) coincide, as can be seen from the relation (7.20), with isobaric ones, and therefore any surface $|\nu| = $ const can be considered free. This allows one to construct new exact solutions where there is a set of homogeneous and stratified layers.

In conclusion, Inogamov gave the solution for epicycloidal waves [30]. It is noteworthy that he derived the expression (7.17) by generalizing the linear solution of this problem, rather than discovering a general formula for Ptolemaic flows (7.4).

References

[1] Lamb, H. (1932). *Hydrodynamics*, 6th edn. (Cambridge University Press, Cambridge).

[2] Batchelor, G.K. (1967). *An Introduction to Fluid Dynamics* (Cambridge University Press, Cambridge).

[3] Milne-Thompson, L.M. (1938). *Theoretical Hydrodynamics* (Macillan & Co. Ltd., London).

[4] Chaplygin, S.A. (1948). A vortex flow flowing over an obstacle in the form of a round semi-cylinder. *Collected Works*, Vol. 2, pp. 537–546 [in Russian] (OGIZ, Moscow-Leningrad).

140   *Analytical Fluid Dynamics in Lagrangian Variables*

[5] Chaplygin, S.A. (1948). A stream flowing around a fence at continuous speeds with the formation of vortices in front and behind the fence. *Collected Works*, Vol. 2, pp. 546–555 [in Russian] (OGIZ, Moscow-Leningrad).

[6] Rankine, W.J. (1958). *Manual of Applied Mechanics* (R. Griffin, London).

[7] Kirchoff, G. (1876). *Vorlesungen uber Mathematische Physic: Mekhanik* (Teubner, Leipzig).

[8] Chaplygin, S.A. (1899). About a pulsating cylindrical vortex. *Proc. Phys. Sect. Mosc. Imp. Soc. Friends Nat. Sci.*, 10(1), 13–22 [in Russian].

[9] Saffman, P.G. (1995). *Vortex Dynamics* (Cambridge University Press, Cambridge).

[10] Moore, D.W. and Saffman, P.G. (1971). Structure of a line vortex in an imposed strain. In: Olsen, J.H., Golgburg, A., and Rogers, M. (eds.), *Aircraft Wake Turbulence and Its Detection*, pp. 339–354 (Springer, Boston, MA).

[11] Kida, S. (1981). Motion of an elliptic vortex in a uniform shear flow. *J. Phys. Soc. Jpn.*, 112, 397–409.

[12] Chaplygin, S.A. (1903). One case of vortex fluid motion. *Proc. Phys. Sect. Mosc. Imp. Soc. Friends Nat. Sci.*, 11, 11–14 [in Russian].

[13] Stuart, J.T. (1967). On finite amplitude oscillations in laminar mixing layer. *J. Fluid Mech.*, 29(3), 417–430.

[14] Shercliff, J.A. (1977). Simple rotational flows. *J. Fluid Mech.*, 82(4), 687–703.

[15] Kaptsov, O.V. (1989). Some classes of two-dimensional vortex flows of an ideal fluid. *J. Appl. Mech. Tech. Phys.*, 30(1), 105–112.

[16] Andreev, V.K., Kaptsov, O.V., Pukhnachov, V.V., and Rodionov, A.A. (1998). *Applications of Group-Theoretic Methods in Hydrodynamics*. MAIA, Vol. 450 (Kluwer Academic Publishers, Dordrecht/ Boston/London).

[17] Cherny, G.G. (1997). Plane steady self-similar vortex flows of an ideal fluid. *Izv. RAN. Mekh. Zhidk. Gaza*, 4, 39–53 [in Russian].

[18] Flierl, G.R., Stern, M.E., and Whitehead, J.A. (1983). The physical significance of modons: laboratory experiments and general integral constraints. *Dyn. Atmos. Oceans*, 7, 233–264.

[19] Abrashkin, A.A. and Yakubovich, E.I. (1984). Two-dimensional vortex flows of an ideal fluid. *Sov. Phys. Dokl.*, 29, 370–371.

[20] Abrashkin, A.A. and Yakubovich, E.I. (1985). Non-stationary vortex flows of an ideal incompressible fluid. *J. Appl. Mech. Tech. Phys.*, 26, 202–208.

[21] Abrashkin, A.A. and Yakubovich, E.I. (2016). Frequence spectrum of exact solutions of the two-dimensional hydrodynamic equations. *Radiophys. Quantum Electron.*, 58, 852–857.

[22] Landau, L.D. and Lifshitz, E.M. (1987). *Fluid Mechanics* (Pergamon Press, Oxford).

[23] Aleman, A. and Constantin, A. (2012). Harmonic maps and ideal fluid flows. *Arch. Rational. Mech. Anal.*, 204, 479–513.

[24] Constantin, O. and Martin, M.J. (2017). A harmonic maps approach to fluid flows. *Math. Ann.*, 369, 1–16.

[25] Martin, M.J. and Tuomela, J. (2019). 2D incompressible Euler equations new explicit solutions. *Discret. Contin. Dyn. Syst. A.*, 39(8), 4547–4563.

[26] Saleva, T. and Tuomela, J. (2021). On the explicit solutions of separation of variables type for the incompressible 2D Euler equations. *J. Math. Fluid Mech.*, 23, 39.

[27] Philips, O.M. (1960). Centrifugal waves. *J. Fluid Mech.*, 7(3), 340–352.

[28] Tsao, S. (1960). On waves on the surface of a liquid under the influence of centrifugal force. *Prikl. Mekh. Tech. Fiz.*, 3, 90–96 [in Russian].

[29] Inogamov, N.A. (1985). A cylindrical analog of trochoidal Gerstner waves. *Fluid Dyn.*, 20(5), 791–796.

[30] Abrashkin, A.A. (1984). Nonlinear azimuthal waves in a centrifuge. *J. Appl. Mech. Tech. Phys.*, 25, 411–415.

Chapter 8

Localized Two-Dimensional Vortices in the External Flow

*"Continuous vortex-free motion is impossible
in an incompressible fluid filling an unbounded
space and subject to the condition that the velocity
becomes zero at infinity."*

— H. Lamb [1]

The problem formulated in the title of this chapter is very simple since, initially, a singly connected region with a known vorticity distribution is specified in an ideal incompressible fluid. It is required to determine the evolution of its boundary.

This problem is solved simply for cylindrical vortices when the streamlines are circles (see Section 3.3.2). Cylindrical vortices are unstable with respect to two-dimensional perturbations in the presence of points where vorticity is extreme [2].

When constructing examples of non-circular regions, the assumption of constant vorticity in the core Ω turns out to be extremely effective. In this case, Eq. (7.1) is satisfied automatically, and therefore the main issue is the binding of internal vortex motion with external potential motion. One approach is based on the assumption that the region differs only slightly from the circle. As Kelvin showed [1], the angular velocity of azimuthal linear waves along the

143

boundary of a circular vortex is equal to

$$\omega_K = \frac{(m-1)\Omega}{2m},$$

where m is the mode order. In turn, Kirchhoff constructed a solution for perturbations of amplitude, according to which a uniformly vortical region in the form of an ellipse uniformly rotates around its center with constant angular velocity (see Section 3.4)

$$\omega = \frac{\Omega AB}{(A+B)^2},$$

where A and B are the semi-axes of an ellipse. A detailed review of Western works on the stability of the Kirchhoff vortex is contained in another study [3] (to which we add the work by Soviet fluid mechanics [4]).

Kirchhoff's exact solution and the approximate one given by Kelvin have long been exceptional examples that gave an idea of the dynamics of two-dimensional vortices. Interest in this issue was revived after the discovery of coherent structures in turbulence and synoptic vortices in the ocean. In relation to the problem under consideration, both analytical and numerical methods have been developed. The structure of uniformly rotating regions of constant vorticity, for which the boundary differs only slightly from the circle, is analyzed by the perturbation method up to the third approximation in terms of a small parameter of the ratio of the perturbation amplitude to the radius [5, 6]. A.G. Petrov proposed a new approach to the study of a uniformly vortical region based on the solution of dynamic Lagrange equations. Petrov [7] studied the equations of a system with two degrees of freedom that determine the rotation and deformation of an ellipse with a fixed center. A special case of their solutions is the Kirchhoff vortex. More general equations that also take into account the ellipse center displacement were obtained in another study [8]. A new one-parameter series of stationary vortex motions has been found. The ellipse rotates around its own center with a constant angular velocity, and the center itself rotates in a circular orbit with the same angular velocity. The major axis of the ellipse is directed toward the circular orbit center. For this series of solutions, the condition of continuous pressure at the boundary of the ellipse is fulfilled only approximately. The error is characterized

Localized Two-Dimensional Vortices in the External Flow 145

by the maximum deviation from zero of the dimensionless pressure discrepancy, which turns out to be quite small in the region of stable stationary motion.

With a uniform distribution of vorticity in a vortex spot, the perturbations themselves are potential, and they can be considered as Kelvin waves. Strongly nonlinear boundary perturbations after the appearance of the paper [9], in which the contour dynamics method was proposed, were studied numerically. The use of this method made it possible to observe non-stationary quasi-return regimes of the region with a cusp and the formation of a vorticity thread as a possible mechanism for vortex destruction. The contour dynamics method also enabled one to significantly expand the range of problems on the dynamics of single vortices and their interaction with each other [9–14]. In contrast to these studies, this chapter will present the analytical results of the description of two-dimensional vortices [15, 16].

8.1.   Ptolemaic Vortices

Consider a singly connected vortex region, potentially streamlined from the outside. Let the expression

$$W = \alpha\nu\exp(i\lambda t) + F(\bar{\nu}); \quad |\nu| \leq 1,$$
$$\nu = \exp(ik\chi), \quad \chi = a + ib, \tag{8.1}$$

where α is a positive constant, specify the Ptolemaic flow in the indicated region. We also assume that on the plane of the Lagrangian variable ν, this region corresponds to the interior of the unit circle, which is equivalent to the condition $b \leq 0$. Equation (8.1) is a special case of Ptolemaic flows (7.4) at $G(\chi) = \exp(ik\chi)$, $\mu = 0$. The shape of the vortex region with this choice can be largely arbitrary, depending on the shape of the function F. The vorticity of this flow is determined by the formula

$$\Omega = \frac{2\lambda}{1 - \alpha^{-2}|F'|^2}. \tag{8.2}$$

For all possible functions F, except for the linear one, the vorticity is minimal at the center of the vortex and increases toward its boundary. The trajectories of fluid particles are circles, each particle having its own rotation center, the position of which is determined by the type of function F.

We now turn to determining the potential flow from the exterior of the vortex. To do this, following John's method (Section 2.7), we write the complex coordinate W of a point in the outer region and the complex velocity V as functions of the complex parameter ν:

$$
\begin{aligned}
W &= \alpha\nu \exp(i\lambda t) + F\left(\frac{1}{\nu}\right), \\
V &= \frac{i\lambda\alpha}{\bar{\nu}} \exp(i\lambda t); \quad |\nu| \geq 1.
\end{aligned}
\tag{8.3}
$$

These relations coincide with the expressions for W and V of the Ptolemaic flow at the boundary of the vortex. Consequently, they satisfy the conditions of continuous velocity and therefore continuous pressure [1] at the boundary.

Equation (8.3) solves the problem since it is a parametric form of writing the desired potential flow. Indeed, it follows that V is an analytical function of the complex conjugate coordinate $\bar{W}$, i.e., the flow is potential. Recall that the variable ν outside the vortex is no longer a Lagrangian coordinate. It coincides with the Lagrangian variable only at the boundary of the vortex $|\nu| = 1$. In this case, we implemented John's method in relation to binding vortex and potential flows.

Note that the law of velocity decay at infinity is the same as for a point vortex. Satisfying this condition was precisely the reason why the shape of the function G and the value of the frequency μ were chosen. In the constructed solution, the function F cannot be arbitrary. It is necessary to fulfill the inequality

$$
|F'| \leq \alpha \quad \text{at} \quad |\nu| \leq 1
\tag{8.4}
$$

so that the Jacobian $D(\chi, \bar{\chi})$ does not become zero inside the vortex (see Eq. (7.5)). It will be shown in the following that the fulfillment of this condition guarantees that the potential velocity field in the exterior of the vortex does not have singularities. It can be seen in Eq. (8.1) that Ptolemaic vortices are generally non-stationary and their shape changes periodically. Figures 8.1–8.3 show the dynamics of the boundary of the vortex region for three different functions F.

Among Ptolemaic vortices, one can extract a family of stationary vortices, i.e., vortices rotating without changing their shape. One of

Localized Two-Dimensional Vortices in the External Flow

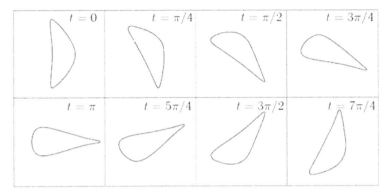

Fig. 8.1. Dynamics of the vortex region in the case $F(\bar{\nu}) = -0.5\bar{\nu} + 0.2\bar{\nu}^2$.

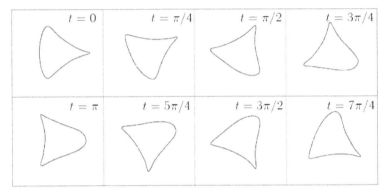

Fig. 8.2. Dynamics of the vortex region in the case $F(\bar{\nu}) = 0.3\bar{\nu}^2 + 0.1\bar{\nu}^3$.

them is the Kirchhoff vortex already considered (Section 3.4). Let us choose the solution to Eq. (8.1) in the following form:

$$W = \alpha \nu \exp(i\lambda t) + \beta \bar{\nu}^n, \qquad (8.5)$$

where β is a positive constant and n is a non-negative integer. Equation (8.5) describes a family of vortices of constant shape, uniformly rotating with angular velocity

$$\omega = \frac{n}{n+1}\lambda.$$

For $n = 1$, we obtain an elliptical vortex. In the case $n \geq 2$, we have hypocycloidal regions with $n + 1$ protrusions. The condition

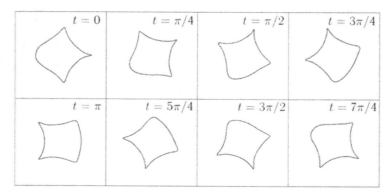

Fig. 8.3. Dynamics of the vortex region in the case $F(\bar{\nu}) = -0.2\bar{\nu}^3 + 0.1\bar{\nu}^4$.

for the absence of self-intersections of the boundary is the inequality $\beta \leq \alpha/n$, which ensures that the velocity field of the potential flow is single valued. The formula for the vorticity of the flow (8.5) is written as follows:

$$\Omega = \frac{2\lambda}{1 - (\beta n/\alpha)^2 |\nu|^{2(n-1)}}. \tag{8.6}$$

For an elliptical vortex, the vorticity is constant, while for other hypocycloidal vortices, it is minimal at the center and increases toward the boundary.

8.2. Nonlinear Kelvin Vortex Waves

Azimuthal waves propagating along the boundary of a circular vortex (Kelvin waves) were observed in a special experiment [17]. The vortex core was created in the space between two rotating disks. The vorticity, which was constant in the main part of the core, dropped sharply in the region of its boundary and was almost zero in the outer region of the flow. During the experiments, it was possible to excite the first few modes. As it turned out, the angular velocity of waves is always less than the velocity of linear waves, and the relative difference between their values can reach 30%.

Ptolemaic vortices are obviously not suitable to explain this fact since their vorticity increases as they approach the boundary. In this

Localized Two-Dimensional Vortices in the External Flow 149

regard, it makes sense to consider weakly nonlinear Kelvin waves with a more general vorticity distribution [18].

8.2.1. *Problem statement and solution method*

Consider a single region of vorticity Ω rotating without changing shape with constant angular velocity in an ideal incompressible fluid. We assume that the flow is two-dimensional and is potential outside the vortex.

It is convenient to study vortex motion in a reference frame rotating with frequency ω. In the reference frame, the flow is stationary, the boundary of the vortex is fixed, and the vorticity inside is equal to $\Omega - 2\omega$. It is convenient to write the equations of two-dimensional fluid dynamics in the following form (see Eq. (3.23)):

$$\frac{D(X,Y)}{D(a,b)} = D_0; \quad \frac{D(X_t,X)}{D(a,b)} + \frac{D(Y_t,Y)}{D(a,b)} = (\Omega - 2\omega)D_0. \quad (8.7)$$

Assume that the boundary of the vortex region on the plane of Lagrangian variables corresponds to the value $b = 0$, and the interior of the core, namely, the half-band $0 \le a \le 2\pi$, $b \ge 0$; $b = \infty$ corresponds to the vortex center. We introduce the complex coordinate of the trajectory $W = X + iY$, $\bar{W} = X - iY$ and the modified Lagrangian coordinates $q = a + \sigma(b)t, b$ (see Chapter 6). The system of Eq. (8.7) will then be rewritten as

$$\frac{D(W,\bar{W})}{D(q,b)} = -2iD_0(b),$$

$$\sigma\frac{D(W_q,\bar{W})}{D(q,b)} - \sigma'|W_q|^2 = (\Omega - 2\omega)D_0. \quad (8.8)$$

The flow vorticity Ω depends only on the coordinate b. A solution inside the vortex is sought in the form

$$W = \exp(i\chi_*) + w(q,b)\exp(i\bar{\chi}_*); \quad \chi_* = q + ib, \quad b \ge 0. \quad (8.9)$$

With allowance for Eq. (8.9), the system (8.8) is solved by the method of successive approximations [18]. The zero approximation is a uniformly vortical circular vortex (the first term in Eq. (8.9)), and the small parameter ε determines the deviation of the vortex boundary

150 *Analytical Fluid Dynamics in Lagrangian Variables*

from the circular one. Unknown functions and flow parameters are expanded into series with respect to this parameter. In particular, the vorticity and angular velocity of rotation are written as

$$\Omega = \Omega_0 + \sum_{n=1}^{\infty} \varepsilon^n \Omega_n(b); \quad \Omega_0 = \text{const},$$

$$\omega = \sum_{n=0}^{\infty} \varepsilon^n \omega_n.$$

The vortical flow (8.9) should be coupled with the potential fluid motion outside the vortex. It is already more convenient to do this in a laboratory system where the boundary of the vortex rotates and the fluid motion is potential outside. The complex coordinate $W^*(q, b, t)$ of the particle trajectory in this reference frame is recalculated with allowance for the known function (8.9) according to the law

$$W^* = [\exp(i\chi_*) + w(q, b) \exp(i\bar{\chi}_*)] \exp(i\omega t), b = \text{Im}\,\chi_* \geq 0. \quad (8.10)$$

The velocity inside the vortex region is determined by the relation

$$V^* = [i\lambda \exp(i\chi_*) + \sigma w_q(q, b) \exp(i\bar{\chi}_*) \\ + i\lambda w \exp(i\bar{\chi}_*)] \exp(i\omega t). \quad (8.11)$$

At the boundary of the vortex, $W^*(q, b = 0, t)$; thus, the velocity distribution $V^*(q, b = 0, t)$ is known. For a potential flow, the complex conjugate velocity $\overline{V^*}$ is a function of the complex coordinate W^*. To satisfy this requirement, it is possible to parametrically extend the functions (8.10) and (8.11) through the boundary $b = 0$ so that the relation

$$\frac{D(\overline{V^*}, W^*)}{D(\chi_*, \bar{\chi}_*)} = 0, \quad \text{Im}\,\chi_* \geq 0$$

is fulfilled. The variable χ_* is a parameter here. The potential continuation method itself is a generalization of John's method (Section 2.7) since $\overline{V^*}$ and W^* can now be functions of two complex quantities.

In a study [18], the formulated problem is solved up to the cubic approximation.

8.2.2. Examples

(a) For potential Kelvin waves, $\Omega_1 = \Omega_2 = 0$. The quantities ω_n, $n = 0, 1, 2$ corresponding to these values are equal to

$$\omega_0 = \frac{(m-1)}{2m}\Omega_0 = \omega_K, \quad \omega_1 = 0, \quad \text{and}$$
$$\omega_2 = -\frac{1}{4}(m-1)\Omega_0, \tag{8.12}$$

respectively, which coincide with the results of other studies [5, 6].

(b) For weakly nonlinear hypocycloidal vortices,

$$\Omega_0 = 2\lambda, \quad \Omega_1 = 0, \quad \text{and} \quad \Omega_2 = 2\lambda(m-1)^2 e^{-2(m-2)b}.$$

We obtain these relations from Eq. (8.6), assuming $\varepsilon = \beta/\alpha$ and $n = m - 1$. Accordingly, for angular velocities, we have

$$\omega_0 = \omega_K, \quad \omega_1 = 0, \quad \text{and} \quad \omega_2 = 0. \tag{8.13}$$

(c) $\Omega_1 = \delta \exp(-\gamma b), \gamma > 0; \quad \Omega_2 = 0$. Depending on the sign of the quantity δ, the vorticity can either increase or decrease as it approaches the boundary of the vortex. Taking into account the form of vorticity Ω_1, we find that

$$\omega_1 = \frac{2\delta(m-1)}{(\gamma+2)(\gamma+2m)}. \tag{8.14}$$

If $\delta > 0$, then the rotation velocity is greater than the velocity $\omega_0 = \omega_K$ of linear waves. If $\delta < 0$, then the rotation velocity is less than ω_0. The latter situation corresponds exactly to another experiment [17]. In the third approximation, we have

$$\omega_2 = -\frac{(m-1)}{4}\Omega_0 - \frac{\delta^2 \gamma (m-1)[\gamma^2 + 2m\gamma + 2m(m-1)]}{\Omega_0(\gamma+2)^2(\gamma+2m)^2(\gamma+m)}, \tag{8.15}$$

which coincides with the third relation in Eq. (8.12) at $\delta = 0$. The ω_2 value is always negative and does not depend on the sign

152   *Analytical Fluid Dynamics in Lagrangian Variables*

of δ. Thus, taking into account the quadratic correction leads to an additional decrease in the angular velocity of wave propagation. Its decrease will become even more significant if Ω_2 is considered negative.

Due to its asymptotic nature, this theoretical analysis cannot be directly compared with the experimental results [17], where the deviations of vorticity from a constant value are not small. Nevertheless, these results indicate a mechanism for reducing the angular velocity of a vortex with "lobes" due to a decrease in vorticity at its periphery, which corresponds to experimental observations.

It is also interesting to note that Eqs. (8.12)–(8.15) are similar to the dispersion relations for Stokes, Gerstner, and Guyon waves, respectively (see Chapters 4–6).

8.3.   Interaction of Two Elliptical Vortices

It is natural to continue the study of oscillations of a single vortex region by analyzing the motion of a system of two distributed vortices [19]. In addition to independent interest, such a problem is also interesting as a model of the "elementary act" of interaction of some isolated vortex pair in a multi-vortex ensemble of vortices.

Consider the motion of two uniformly vortical regions, the characteristic sizes of which L_1 and L_2 are much smaller than the distance L between them. The latter circumstance makes it possible, instead of the self-consistent motion of two vortices, to study the dynamics of a single distributed vortex in a given external non-stationary field created by another vortex.

The dynamics of two plane vortices can be fully studied assuming they are points. When considering the motion of vortices in the next approximation by a small parameter of the ratio of the vortex size to the distance between them, it should be assumed that the vorticity centers of distributed vortices, W_1 and W_2 $(W = X + iY)$, move as those of point vortices of the same intensity. They uniformly rotate around the common center of vorticity with a constant angular velocity equal to

$$\omega_0 = \frac{\kappa_1 + \kappa_2}{2\pi L^2},$$

Localized Two-Dimensional Vortices in the External Flow 153

where κ_1 and κ_2 are the intensities of the first and second vortices, respectively.

The potential velocity field near one of the vortices, e.g., the first one, can be written as follows:

$$\bar{V}(W,t) = \bar{V}_1(W,t) + \frac{\Omega_2}{2\pi i} \iint_{\Pi_2} \frac{d\sigma(W',\overline{W'})}{W - W'}, \qquad (8.16)$$

where $\bar{V}_1(W,t)$ is the velocity field created by the first vortex (the bar is the sign of complex conjugation) and $d\sigma(W',\overline{W'})$ is an element of the area of the second vortex Π_2, the vorticity Ω_2 of which is assumed to be constant.

We will count the integration variable from the center of second vortex. Equation (8.16) in this case will take the form

$$\bar{V}(W,t) = \bar{V}_1(W,t) + \frac{\Omega_2}{2\pi i} \iint_{\Pi_2} \frac{d\sigma(W'',\overline{W''})}{W - W_2 - W''}. \qquad (8.17)$$

If we switch to a reference frame where the center of the first vortex is at rest (Eq. (8.17), we should replace $W \to W + W_1$), and expand the integrand function in a series with respect to the parameter $(W - W'')/(W_2 - W_1)$, limiting ourselves to terms linear in W and W', then the velocity field near the first vortex will be written as

$$\bar{V}(W,t) = \bar{V}_1(W,t) - \frac{\kappa_2}{2\pi i} \frac{W}{L^2} e^{-2i\omega_0 t}. \qquad (8.18)$$

Here, it is assumed for simplicity that at the initial moment the centers of the vortices lie on a real line and, therefore,

$$W_2 - W_1 = L \exp(i\omega_o t).$$

Note that in the considered approximation, the velocity distribution created by the second vortex near the first one does not depend on the shape of the second vortex, since the integral

$$\iint_{\Pi_2} W'' d\sigma(W'',\overline{W''})$$

is equal to zero (this follows from the definition of the vortex center).

Thus, the effect of one of the vortices on the dynamics of the other is equivalent to the action of a linear (in complex coordinate) and time-harmonic deformation field near the latter. In the case of a finite set of vortices, the deformation field acting on some selected vortex will be proportional to the complex coordinate, unless the vortex sizes are small compared to the distances to the other vortices. The dependence of the deformation field on time (which is determined by the solution of the zero approximation, when vortices are considered points) can be more complex than the harmonic one. In general, the velocity field near the selected vortex will have the form

$$\bar{V}(W,t) = \bar{V}_1(W,t) + 2i\varepsilon(t)W, \tag{8.19}$$

where $\varepsilon(t)$ is a known function of time.

8.3.1. *Generalized Kirchhoff vortices*

For the purposes of this study, it is convenient to consider a separate problem on the dynamics of a single vortex located in an infinite fluid in a uniform deformation non-stationary external field. In this case, the expression for the velocity field (8.19) will already be valid in the entire region outside the vortex, not only near its boundary. Note that the velocity of the potential flow outside the vortex near an infinitely distant point is equal to the second term on the right-hand side of Eq. (8.19). In relation to the vortex interaction problem, solving the latter will make sense only inside the vortex and near its boundary.

Let us first consider the motion in the vortex. We will seek a solution to Eqs. (3.11) and (3.12) in the following form [16]:

$$W = \chi g(t) + \bar{\chi} f(t).$$

It follows from the continuity equation that

$$|g|^2 - |f|^2 = D_0 = \text{const.} \tag{8.20}$$

From the momentum equation, we have

$$\dot{g}g - \dot{f}f = \frac{i\Omega D_0}{2}, \tag{8.21}$$

Localized Two-Dimensional Vortices in the External Flow 155

where Ω is a constant vorticity and the dot indicates differentiation in time. Solving Eqs. (8.20) and (8.21), we obtain the form of the desired solution

$$W = \chi|g|\exp(i\varphi) + \bar{\chi}\sqrt{|g|^2 - D_0}\exp(i\psi), \qquad (8.22)$$

where $\varphi(t)$ and $\psi(t)$ are the arguments of the functions g and f connected by the relation

$$\dot{\psi} = \frac{2\dot{\varphi}|g|^2 - \Omega D_0}{2(|g|^2 - D_0)}. \qquad (8.23)$$

Thus, Eqs. (8.22) and (8.23) describe a class of non-stationary flows with constant vorticity. In what follows, we assume that these relations specify the motion inside the vortex. We also assume that in the plane of the Lagrangian variable, the vortex flow region corresponds to the interior of the unit circle.

From the form of solution (8.22), it is easy to conclude that the boundary of the vortex ($|\chi| = 1$) will be an ellipse. As a result, it is convenient to introduce new functions $A(t)$ and $B(t)$, which have the meaning of semi-axes of an ellipse and are connected to the functions $|g|$ and D_0 by the relations

$$AB = D_0 \quad \text{and} \quad |g|^2 = \frac{1}{4}(A + B)^2. \qquad (8.24)$$

In this case, Eq. (8.22) will take the form

$$W = \frac{1}{2}\chi(A + B)\exp(i\varphi) + \frac{1}{2}\bar{\chi}(A - B)\exp(i\psi), \quad A \geq B. \qquad (8.25)$$

In addition to deformation (A and B are functions of time; the constancy of their product means that the area of the vortex is constant), the vortex rotates with instantaneous angular velocity

$$w(t) = \frac{1}{2}(\dot{\varphi} + \dot{\psi}) = \frac{\dot{\varphi}(A^2 + B^2) - \Omega AB}{(A - B)^2}. \qquad (8.26)$$

Thus, the solution (8.25) describes the most general case of vortex motion (A, B, and w are arbitrary functions of time), in which the vortex remains an ellipse. The form of these functions is determined by the flow outside the vortex. In particular, if we assume A and B

156   *Analytical Fluid Dynamics in Lagrangian Variables*

to be independent of time and $\dot{\varphi} = \lambda = \text{const}$ and $\dot{\psi} = 0$, the solution (8.25) will describe the Kirchhoff vortex (see Section 3.4), and if we assume that $\dot{\varphi} = \lambda = \text{const}$ and $\dot{\psi} = \mu = \text{const}$, it will describe a family of elliptical vortices rotating without changing shape. A special case of it (for $\lambda = -\mu$) is the stationary Saffman and Moore vortex [20, 21]. The solution (8.25) also describes the Chaplygin [22] and Kida [23] vortices, i.e., non-stationary elliptical vortices located, respectively, in the shear flow and the flow specified by the relation

$$\bar{V} = \delta W - i\gamma \bar{W}.$$

where δ and γ are real constants. All of these vortices can be called generalized Kirchhoff vortices. Their generality is that they are described by Eq. (8.25).

We now consider the problem of binding the motion (8.25) with an external potential flow having at infinity the asymptotic form $\psi_\infty = \varepsilon(t)[x^2 - y^2]$, or (for the velocity field) $\bar{V}_\infty = 2i\varepsilon(t)W$. According to our choice, the vortex motion on the Lagrangian plane corresponds to the interior of the unit circle. The velocity inside the vortex is given by the expression

$$V = \frac{1}{2}\chi[\dot{A} + \dot{B} + i\dot{\varphi}(A + B)]\exp(i\varphi)$$

$$+ \frac{1}{2}\bar{\chi}[\dot{A} - \dot{B} + i\dot{\psi}(A - B)]\exp(i\psi), \quad |\chi| \leq 1.$$

For a potential flow, the complex conjugate velocity $\bar{V}$ is a function of the complex coordinate W. Therefore, the relations

$$W = \frac{1}{2}(A + B)\chi\exp(i\varphi) + \frac{1}{2}(A - B)\frac{1}{\chi}\exp(i\psi), \quad |\chi| \geq 1,$$

$$\bar{V} = \frac{1}{2}[\dot{A} + \dot{B} - i\dot{\varphi}(A + B)]\frac{1}{\chi}\exp(-i\varphi) \qquad (8.27)$$

$$+ \frac{1}{2}[\dot{A} - \dot{B} - i\dot{\psi}(A - B)]\chi\exp(-i\psi),$$

where χ serves as a parameter (and coincides with the complex Lagrangian coordinate only at the vortex boundary), solve the problem of the potential continuation of vortex motion (8.25) with continuous velocity at the bonding boundary (John's method, Section 2.7).

Localized Two-Dimensional Vortices in the External Flow 157

The representation of $\bar{V}(W,t)$ is unique since $W_\chi \neq 0$ is outside the vortex.

Near an infinitely distant point, the velocity field is written as follows:

$$\bar{V}_\infty = \frac{\dot{A} - \dot{B} - i\dot{\psi}(A-B)}{A+B} W \exp[-i(\varphi + \psi)],$$

and therefore the parameters of the ellipse should satisfy the condition

$$\frac{\dot{A} - \dot{B} - i\dot{\psi}(A-B)}{A+B} \exp[-i(\varphi + \psi)] = 2i\varepsilon(t). \tag{8.28}$$

Together with equalities (8.23) and (8.24), this condition forms a system of equations for determining four unknown functions, namely, A, B, φ, and ψ.

8.3.2. *Time-harmonic deformation field*

Let us rewrite Eq. (8.28) with allowance for Eq. (8.18):

$$\frac{\dot{A} - \dot{B} - i\dot{\psi}(A-B)}{A+B} \exp[-i(\varphi + \psi)] = \frac{i\kappa_2}{2\pi} \frac{W}{L^2} \exp(-2i\omega_0 t). \tag{8.29}$$

We now add two more relations to this complex equation, which follow from Eqs. (8.23) and (8.24):

$$(A-B)^2\dot{\psi} = (A+B)^2\dot{\varphi} - 2\Omega AB, \quad AB = D_0. \tag{8.30}$$

Let us introduce new variables:

$$\theta = \frac{A}{B}; \quad \sigma = \varphi + \psi - 2\omega_0 t.$$

In them, Eqs. (8.29) and (8.30) will be rewritten as follows:

$$\dot{\theta} = -2\Gamma_2\theta\sin\sigma, \quad \Gamma_2 = \frac{\kappa_2}{2\pi L^2}, \tag{8.31}$$

$$\dot{\sigma} = -2\omega_0 - 2\frac{\Gamma_2(\theta^2+1)}{\theta^2-1}\cos\sigma + \frac{2\Omega\theta}{(\theta+1)^2}. \tag{8.32}$$

Thus, the dynamics of a single vortex in the field created by another vortex is described by a system of two autonomous equations of the first order. It will be shown in the following that the system is reduced to quadratures. But first, it is convenient to consider its properties using the phase plane. We study the stationary solutions of the systems (8.31) and (8.32). For the equilibrium state, which corresponds to the value $\sigma = 0$, the semi-axis ratio is determined by the following equation:

$$\kappa_1(\theta + 1)^2(\theta - 1) - 2\pi L^2\Omega\theta(\theta - 1) + 2\kappa_2\theta^2(\theta + 1) = 0. \qquad (8.33)$$

This is a cubic equation with respect to θ, and therefore it has three roots. Let us find them. To do this, we write Eq. (8.33) in the following form:

$$\frac{(\theta + 1)\left[\theta^2\left(1 + 2\frac{\kappa_2}{\kappa_1}\right)\right]}{\theta(\theta - 1)} = \frac{2\pi L^2\Omega}{\kappa_1}. \qquad (8.34)$$

The right-hand side of the equality is much larger than unity. Indeed, since the circulation is equal to the product of vorticity by the area of the vortex, an estimate

$$\frac{2\pi L^2\Omega}{\kappa_1} = \frac{2\pi L^2}{S} \approx \left(\frac{L}{L_1}\right)^2 \gg 1$$

is valid for this constant. Examining the location of the roots of Eq. (8.34) using the graphical method, it is not difficult to show that if the intensity of vortices is of the same sign $(\kappa_2/\kappa_1 > 0)$, then the following relations are valid for the roots of the equation:

$$0 < \theta_1 \ll 1, \quad \theta_2 \geq 1, \quad \text{and} \quad \theta_3 \gg 1.$$

For vortices with intensities of different signs $(\kappa_2/\kappa_1 < 0)$, these relations take the form

$$0 < \theta_1 \ll 1, \quad \theta_2 \leq 1, \quad \text{and} \quad \theta_3 \gg 1.$$

The study of the equilibrium position corresponding to the value $\sigma = \pi$ can be easily reduced to the case just considered since the form of the system (8.31) and (8.32) is invariant with respect to the replacements

$$\theta \to \theta^{-1} \quad \text{and} \quad \sigma \to \sigma + \pi.$$

As a result, the $\sigma = \pi$ value corresponds to the semi-axis ratios $\theta_1^{-1}, \theta_2^{-1}$, and θ_3^{-1} in the equilibrium state. Due to the choice of the

Localized Two-Dimensional Vortices in the External Flow 159

form of solution (8.25), θ exceeds unity. For "stationary" values of θ, significantly exceeding unity, the assumption of the smallness of vortices in comparison with the distance between them is violated. Indeed, from Eq. (8.34) for $\theta \gg 1$, we have $\theta = A/B \approx 2\pi L^2 S^{-1} \gg 1$; since $AB = S/\pi$, we find that the semi-axis A is of the order L.

Thus, the stationary solutions of systems (8.31) and (8.32), satisfying the conditions of the problem under consideration, correspond to the values $\sigma = 0$ and $\theta = \theta_2$ in the case of $\kappa_2/\kappa_1 > 0$ and to the values $\sigma = \pi$ and $\theta = \theta_2^{-1}$ in the case $\kappa_2/\kappa_1 < 0$. In both cases, the semi-axis ratio slightly exceeds unity, i.e., stationary vortices differ only slightly from circular ones (non-stationary vortices, naturally, can be significantly elliptical if the condition of small semi-axes compared to L is met).

The magnitude of the angular velocity of rotation of the vortex around its own center of inertia for a stationary state, as follows from Eqs. (8.26) and (8.32), is equal to ω_0; therefore, the vortex does not change its orientation relative to the straight line connecting the centers of vortices (the centers of vortices move as point ones of the same intensity as the distributed vortices under consideration). The orientation of the ellipse relative to this line can be determined by considering its position at the initial moment of time when the coordinates of the vortex centers are real. Since, in the case of vortices with intensities of the same sign, the value $\theta \geq 1$ at $\sigma = 0$, the major axis of the ellipse is directed along the line that connects the centers of the vortices. In the case of vortices with different sign intensities, the value $\theta \geq 1$ corresponds to $\sigma = \pi$, and therefore the minor axis lies on the indicated line.

Exactly the same consideration can be performed for the second vortex (in all ratios, the indices 1 and 2 will be reversed). Thus, in the first approximation, the stationary state is one in which ellipses of constant size rotate around the vorticity center with a constant orientation relative to the line connecting their centers (Fig. 8.4).

In the case where the orientation of the elliptical vortex regions does not coincide with the orientation of the vortices corresponding to the stationary state, the ellipses will rotate non-uniformly and deform. For a qualitative analysis of this motion, it is convenient to employ the phase plane. Let us first define the types of equilibrium

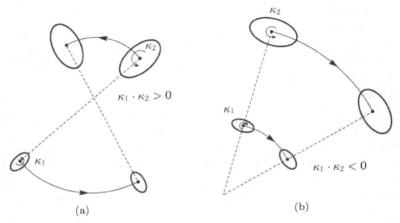

Fig. 8.4. Orientation and trajectories of motion of vortices of constant shape.

states. Upon linearization, systems (8.31)–(8.32) will take the form

$$\dot{\tilde{\theta}} = -2\Gamma_2 \theta_0 \tilde{\sigma} \cos \sigma_0, \qquad (8.35)$$

$$\dot{\tilde{\sigma}} = \left[\frac{8\theta_0 \Gamma_2}{(\theta_0 - 1)^2} \cos \sigma_0 - \frac{2\Omega(\theta_0 - 1)}{(\theta_0 + 1)^2} \right] \tilde{\theta}, \qquad (8.36)$$

where θ_0 and σ_0 are the values of the quantities in the equilibrium state, and $\tilde{\theta}$ and $\tilde{\sigma}$ are deviations from them. By finding the roots of the characteristic equation for this system, we conclude that the equilibrium states close to unity of the θ values are the centers. The equilibrium states corresponding to θ values significantly exceeding unity are the saddle for $\sigma_0 = 0$ and the center in the case $\sigma_0 = \pi$.

Figure 8.5 shows the phase plane for the cases $\kappa_1 \kappa_2 > 0$ and $\kappa_2 < 0$. The orientation of the ellipse relative to the straight line connecting the centers of the vortices is determined by the σ value. Indeed, Eq. (8.25) can be rewritten as follows:

$$W = \frac{1}{2} \left[(A+B)\chi \exp \frac{i(\varphi - \psi)}{2} + (A-B)\bar{\chi} \exp \frac{i(\psi - \varphi)}{2} \right]$$

$$\times \exp \frac{i(\varphi + \psi)}{2}.$$

The expression in square brackets describes an ellipse, the major axis of which lies on the horizontal axis. The exponential factor

Localized Two-Dimensional Vortices in the External Flow 161

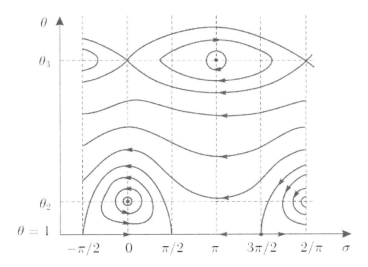

Fig. 8.5. Phase plane for systems (8.31)–(8.32).

outside the brackets describes the rotation of the ellipse around the center. The angle of rotation of the major axis relative to the horizontal is

$$\frac{1}{2}(\varphi + \psi) = \frac{\sigma}{2} + \omega_0 t.$$

The time-linear term in this equation corresponds to the uniform rotation of the ellipse, in which its orientation relative to the straight line connecting the centers of vortices does not change (the centers of vortices rotate with the same angular velocity relative to the center of vorticity). Therefore, the term $\sigma/2$ determines the ellipse rotation relative to the straight line connecting the centers of the vortices.

It follows that if at the initial moment the vortex is rotated by an angle less than $\pi/4$ with respect to its stationary position, it oscillates relative to this position (closed trajectories on the phase plane). In the case of small deviations from the equilibrium state, the oscillation eigenfrequency ω' of the vortex is equal to (see Eqs. (8.35) and (8.36))

$$\omega' = \frac{4\theta_0 \Gamma_2}{\theta_0 - 1}.$$

The oscillation eigenfrequency of the vortex significantly exceeds the rotational frequency of the vortices since

$$\frac{\omega'}{\omega_0} = \frac{4\theta_0\kappa_2}{(\kappa_1 + \kappa_2)(\theta_0 - 1)},$$

and the θ_0 value is close to unity. When the deviation from the stationary position exceeds the angle $\pi/4$, e.g., when the vortices have intensities of the same sign with their minor axes facing each other at the initial time, the vortices tend to rotate so that their major semi-axes face each other (such motion on the phase plane corresponds to open trajectories). If the vortex is circular at the initial time, then, as follows from Fig. 8.5, it will stretch in the direction of the straight line connecting the centers of the vortices. This conclusion is consistent with the results of the numerical experiment [9].

For vortices with different sign intensities, the phase plane has the same form, but the whole picture should be shifted by π over σ. With this observation in mind, it is easy to reformulate the above-mentioned results on the vortex dynamics. In particular, the initially circular vortices will stretch across the straight line connecting the centers, etc.

To complete the analysis of the properties of systems (8.31)–(8.32), we give the form of the exact solution. It has the first integral of the following form:

$$\frac{2(\theta^2 - 1)\Gamma_2 \cos \sigma}{\theta} = 2\Omega \ln \frac{(\theta + 1)^2}{\theta} - 2\omega_0 \frac{\theta^2 + 1}{\theta} + C_0.$$

Here, the constant C_0 is determined by the initial conditions. Using this representation, it is possible to find an implicit expression for the function $\theta(t)$, which is written as

$$t - t_0 = -\frac{1}{2\Gamma_2} \int_{\theta(t_0)}^{\theta} \left[1 - \frac{\theta^2}{4\Gamma_2^2(\theta - 1)^2}\right.$$

$$\left. \times \left(2\Omega \ln \frac{(\theta + 1)^2}{\theta} - 2\omega_0 \frac{\theta^2 + 1}{\theta}\right)^2\right]^{-1/2} \frac{d\theta}{\theta}.$$

By specifying the value of the constants included in this relation, it is possible to determine the time evolution of the ellipse semi-axis ratio in each specific case.

8.4. Non-Lagrangian Approaches

The idea of considering the non-stationary dynamics of distributed vortices assuming that their sizes L_i are small compared to distances between any two vortices of a given L_{jk} system was first proposed in a different study [24]. The quantity $\Delta_i = L_i/\min L_{ik}$ serves as a small parameter for the i-th vortex. The solution obtained assuming that the vortices are point ones serves as the zero approximation. The vortices can be assumed elliptical in the first approximation in terms of Δ_i. The dynamics of each vortex depends on the vorticity center coordinates, the ellipse semi-axis ratio, and the angle determining the direction of its main axis. In this work, examples of the interaction of two initially circular regions are numerically calculated. The calculation was performed in two ways, namely, using an "elliptical model" and within the framework of exact equations (employing the contour dynamics method). Based on these calculations, the limits of the possible application of the approximate model are indicated.

In a study [25], the Hamiltonian formulation is given and the basic conservation laws for the model of small elliptical vortices are found.

Essentially, this model is the generalization of the example previously considered by Moore and Saffman [21, 26] in relation to a chain of identical vortices. These authors, assuming that the size of the vortex is small compared to the distance to the nearest neighbors, considered the problem of the parameters of a single elliptical vortex located in a stationary uniform deformation field. Our study described in the previous section differs from one of Moore and Saffman's studies [21] in that the deformation field is already assumed to be non-stationary. The initial idea of Moore and Saffman also gave rise to a whole range of studies (see, e.g., [27–29]) related to examining the dynamics of an elliptical vortex in an external deformation field, the properties of which are chosen without reference to a specific physical situation.

The interaction of the two vortex regions has intensely been explored using numerical methods. This approach permits one to study in detail the most interesting moments of the evolution of a pair of vortices, when one vortex, e.g., twists around another or the vortex shroud is "pulled out" by the neighbor vortex. Among the works devoted to the analysis of this process, some papers [30, 31] contain, among other things, a very detailed bibliography on this topic.

References

[1] Lamb, H. (1932). *Hydrodynamics*, 6th ed. (Cambridge University Press, Cambridge).

[2] Betchov, R. and Criminale, W. (1967). *Stability of Parallel Flows* (Academic Press, New York).

[3] Mitchell, T.B. and Rossii, L.F. (2008). The evolution of Kirchhoff elliptic vortices. *Phys. Fluids*, 20, 054103.

[4] Vladimirov, V.A. and Il'in, K.I. (1988). Three-dimensional instability of an elliptic Kirchhoff vortex. *Fluid Dynamics*, 23(3), pp. 356–360.

[5] Su, C.H. (1979). Motion of fluid with constant vorticity in a single connected region. *Phys. Fluids*, 22(10), pp. 2032–2033.

[6] Mindlin, I.M. (1984). On vorticity induced waves in a homogeneous incompressible fluid. *J. Appl. Math. Mech.*, 48(5), pp. 550–555.

[7] Petrov, A.G. (1980). About the movement of the Gulf Stream rings. *Okeanologiya*, 20(6), pp. 965–973 [in Russian].

[8] Perepelkin, V.V. and Petrov, A.G. (1983). Dynamics of an elliptic vortex. *Fluid Dynamics*, 18(4), pp. 539–544.

[9] Dim, G.S. and Zabusky, N.J. (1978). Stationary "V-states", interactions, recurrence and breaking. In: K. Longren and A. Scott, (Eds.), *Solitons in Action* (Academic Press, New York).

[10] Pullin, D.L. (1992). Contour dynamics methods. *Annu. Rev. Fluid Mech.*, 24, pp. 89–115.

[11] Crowdy, D.G. and Surana, A. (2007). Contour dynamics in complex domains. *J. Fluid Mech.*, 593, pp. 235–254.

[12] Dritschel, D.G. (1989). Contour dynamics and contour surgery: Numerical algorithms for extended, high-resolution modelling of vortex dynamics in two-dimensional, inviscid, incompressible flows. *Comput. Phys. Rep.*, 10(3), pp. 77–146.

[13] Wu, H.M., Overman, E.A. and. Zabusky, N.J. (1984). Steady-state solutions of the Euler equations in two dimensions: Rotating and translating V-states with limit cases: 1. numerical algorithms and results, *J. Comput. Phys.*, 53(1), pp. 42–71.

[14] Zabusky, N.J., Hughes, M.H. and Roberts, K.V. (1979). Contour dynamics for the Euler equations in two dimensions, *J. Comput. Phys.*, 30(1), pp. 96–106.

[15] Abrashkin, A.A. and Yakubovich, E.I. (1984). Two-dimensional vortex flows of an ideal fluid. *Soviet Phys. Dokl.*, 29, pp. 370–371.

Localized Two-Dimensional Vortices in the External Flow   165

[16]   Abrashkin, A.A. and Yakubovich, E.I. (1985). Non-stationary vortex flows of an ideal incompressible fluid. *J. Appl. Mech. Tech. Phys.* 26, pp. 202–208.

[17]   Akhmetov, A.G. and Tarasov, V.F. (1986). Structure and evolution of vortex cores. *J. Appl. Mech. Tech. Phys*, 27(5), pp. 68–73.

[18]   Abrashkin, A.A. and Zen'kovich, D.A. (1997). Nonlinear Kelvin waves at the boundary of a cylindrical vortex. *Mekh. Zhidk. Gaza,* 5, pp. 62–70 [in Russian].

[19]   Abrashkin, A.A. (1987). On the theory of interaction of two plane vortices. *Fluid Dynamics*, 22(1), pp. 53–59.

[20]   Saffman, P.G. (1992). *Vortex Dynamics* (Cambridge University Press, Cambridge).

[21]   Moore, D.W. and Saffman, P.G. (1971). Structure of a line vortex in an imposed strain. In: Olsen, J.H., Goldburg, A. and Rogers, M., (Eds.), *Aircraft Wake Turbulence and its Detection,* pp. 339–354.

[22]   Chaplygin, S.A. (1899). About a pulsating cylindrical vortex. *Proc. of the Physical Section of the Moscow Imperial Society of Friends of Natural Science,* 10(1), pp. 13–22 [in Russian].

[23]   Kida, S. (1981). Motion of an elliptic vortex in a uniform shear flow. *J. Phys. Soc. Japan.*, 112, pp. 397–409.

[24]   Melander, M.V., Styszek, A.S. and Zabusky, N.J. (1984). Elliptically desingularized vortex model for the two-dimensional Euler equations. *Phys. Rev. Lett.*, 53(13), pp. 1222–1225.

[25]   Brutyan, M.A. and Krapivsky, P.L. (1988). Hamiltonian formulation and fundamental conservation laws for a model of small elliptical vortices. *J. Appl. Math. Mech.*, 52(1), pp. 133–136.

[26]   Moore, D.W. and Saffman, P.G. (1975). The density of organized vortices in a turbulent mixing layer. *Fluid Mech.*, 69, pp. 465–473.

[27]   Neu, J.C. (1984). The dynamics of a columnar vortex in an imposed strain. *Phys. Fluids*, 27(10), pp. 2397–2402.

[28]   Polvani, L.M. and Wisdom, J. (1990). Chaotic Lagrangian trajectories around an elliptical vortex path embedded in a constant and uniform background shear flow. *Phys. Fluids A*, 2(2), pp. 123–126.

[29]   Dhanak, M.R. and Marshall, M.P. (1993). Motion of an elliptical vortex under applied periodic strain. *Phys. Fluids A*, 5(5), pp. 1224–1230.

[30]   Dritshel, D.G. and Waugh, D.W. (1992). Quantification of the inelastic interaction of the unequal vortices in two-dimensional vortex dynamics. *Phys. Fluids A*, 4(8), pp. 1737–1744.

[31]   Waugh, D.W. (1992). The efficiency of symmetric vortex merger. *Phys. Fluids A.*, 4(8), pp. 1745–1758.

Chapter 9

Spatial Vortex Flows

> "The vortices, like beasts in fury,
> Are swarming on you in vain,
> Your heavy depths
> Lose interest in their fight."

> — Innokenty Annensky

In the analytical theory of three-dimensional vortex flows of an ideal incompressible fluid, two approaches can be distinguished as the main ones.

The first approach is related to the study of steady axisymmetric swirling flows. To analyze such motion, as in the planar case, it is possible to introduce a stream function that satisfies the equation [1]

$$\psi_{zz} + \psi_{rr} - \frac{1}{r}\psi_r = r^2 \frac{dH}{d\psi} - C\frac{dC}{d\psi}, \tag{9.1}$$

where r, z, and φ are cylindrical coordinates and H and C are arbitrary functions of ψ. The quantity H has the meaning of the Bernoulli constant, and C is proportional to the azimuthal velocity $V_\varphi(r,z)$ and characterizes the twist of the flow. In plasma physics, Eq. (9.1) is called the Grad–Shafranov equation [2–4]. The classical solution for the Hill vortex [5] is obtained from Eq. (9.1) if we assume

$$H = -A\psi \quad \text{and} \quad C = 0$$

to be equal inside a sphere with a radius r_0 moving relative to a fluid at rest at infinity with the velocity $2Ar_0^2/15$, and zero outside the sphere. Hicks [6] and later Moffat [7] considered a more complex

168   *Analytical Fluid Dynamics in Lagrangian Variables*

flow for which both functions inside a certain sphere are linear in ψ. Such a vortex motion is already swirling but still can be bonded with an external potential flow that is homogeneous at infinity. A similar choice of functions H and C was analyzed in a study [8]. The exact solution 9.1 is also found for the quadratic function H and the linear function C [9]. The solution describes solitary toroidal vortices; however, in these solutions the derivative ψ_z undergoes a jump on the axis. In another study [10], the properties of Eq. (9.1) are analyzed for the case where

$$H = \frac{1}{2}A\psi^2 \quad \text{and} \quad C^2 = 2B \int \psi \ln |\psi| d\psi.$$

Here, B is the constant. This choice of arbitrary functions also makes it possible to describe a set of toroidal structures (see [11]).

The second main approach is related to the study of the dynamics of fluid ellipsoids (see Section 3.6).

A number of results, which can be obtained on the basis of a matrix representation of the equations of fluid dynamics [12, 13], are presented in the following.

9.1. Complex form of Matrix Equations of Fluid Dynamics

Let us introduce into consideration a matrix L_{ik} such that

$$L_{ik} = \frac{\partial W_i}{\partial \chi_k}, \tag{9.2}$$

where W_i and χ_k are determined by the formulas

$$\chi_1 = \frac{a + ib}{\sqrt{2}}, \quad \chi_2 = \frac{a - ib}{\sqrt{2}}, \quad \chi_3 = c,$$

$$W_1 = \frac{X + iY}{\sqrt{2}}, \quad W_2 = \frac{X - iY}{\sqrt{2}}, \quad W_3 = Z.$$

In the new notation, Eqs. (1.26) and (1.27) will be rewritten as

$$\hat{L}_t^*\hat{L} - \hat{L}^*\hat{L}_t = \hat{S}_c, \tag{9.3}$$

$$\frac{\partial L_{ik}}{\partial \chi_n} = \frac{\partial L_{in}}{\partial \chi_k}, \tag{9.4}$$

where * is the sign of the Hermitian conjugation and $\hat{S}_c$ is a complex matrix composed of Cauchy invariants:

$$\hat{S}_c = \begin{pmatrix} -iS_3 & 0 & (-S_2+iS_1)/\sqrt{2} \\ 0 & iS_3 & (-S_2-iS_1)/\sqrt{2} \\ (S_2+iS_1)/\sqrt{2} & (S_2-iS_1)/\sqrt{2} & 0 \end{pmatrix}.$$

The system of Eqs. (9.3) and (9.4) should be supplemented with the continuity equation

$$\det \hat{L} = \det \hat{L}_0, \tag{9.5}$$

and it follows from the representation of the matrix $\hat{L}$ (9.2) that there are also constraints on the choice of its elements:

$$L_{22} = \bar{L}_{11}, \quad L_{12} = \bar{L}_{21}, \quad L_{23} = \bar{L}_{13}, \quad L_{31} = \bar{L}_{32}, \quad \mathrm{Im}\, L_{33} = 0, \tag{9.6}$$

where the bar denotes complex conjugation.

Conditions (9.3)–(9.6) represent a complex form of matrix equations of fluid dynamics [12, 13]. Let us test this approach using the example of Ptolemaic flows. For them, $S_1 = S_2 = 0$, the invariant S_3 is (see Eqs. (1.32) and (7.6))

$$S_3 = 2[\lambda|G'|^2 - \mu|F'|^2],$$

and the matrix $\hat{L}$ is written as follows:

$$\hat{L} = \begin{pmatrix} e^{\frac{i}{2}(\lambda+\mu)t} & 0 \\ 0 & e^{-\frac{i}{2}(\lambda+\mu)t} \end{pmatrix} \begin{pmatrix} G'(\chi_1) & F'(\chi_2) \\ \bar{F}' & \bar{G}' \end{pmatrix}$$

$$\times \begin{pmatrix} e^{\frac{i}{2}(\lambda-\mu)t} & 0 \\ 0 & e^{-\frac{i}{2}(\lambda-\mu)t} \end{pmatrix}. \tag{9.7}$$

Note that a kind of double separation of variables occurred in the indicated representation. Firstly, the "temporal" and "spatial" matrices were separated, and secondly, within the "coordinate" matrix, each column depends on only one complex coordinate.

9.2. Generalized Ptolemaic Flows

We will seek a solution to Eqs. (9.3)–(9.6) in the form

$$\hat{L} = \mathrm{diag}\{e^{-i\gamma t}, e^{i\gamma t}, 1\}\hat{L}_0 \, \mathrm{diag}\{e^{-i\omega t}, e^{i\omega t}, 1\}, \tag{9.8}$$

where the matrix $\hat{L}_0$ is specified as

$$\hat{L}_0 = \begin{pmatrix} L_{11}(\chi_1) & L_{12}(\chi_2) & L_{13}(\chi_3) \\ L_{21}(\chi_1) & L_{22}(\chi_2) & L_{23}(\chi_3) \\ L_{31}(\chi_1) & L_{32}(\chi_2) & L_{33}(\chi_3) \end{pmatrix}. \tag{9.9}$$

This form of the solutions generalizes the expression (9.7); therefore, these ones are naturally called generalized Ptolemaic in what follows.

The matrix expression (9.8) satisfies the continuity Eq. (9.5). The gradient condition (9.4) is fulfilled for the matrix $\hat{L}_0$ and is preserved at any moment for the matrix $\hat{L}$ as well due to the diagonality of the matrices that are only time-dependent. Therefore, it remains to be clarified under what constraints on the $\hat{L}_0$ elements the momentum Eq. (9.3) is satisfied.

To do this, we write the expression for the matrix $\hat{L}$ in its explicit form

$$\hat{L} = \begin{pmatrix} L_{11}e^{-i(\omega+\gamma)t} & L_{12}e^{i(\omega-\gamma)t} & L_{13}e^{-i\gamma t} \\ L_{21}e^{i(\omega-\gamma)t} & L_{22}e^{i(\omega+\gamma)t} & L_{23}e^{i\gamma t} \\ L_{31}e^{-i\omega t} & L_{32}e^{i\omega t} & L_{33} \end{pmatrix}$$

and substitute it into Eq. (9.3), from which it follows that

$$S_1 = S_2 = 0,$$
$$S_3 = -2[(\omega+\gamma)|L_{11}|^2 + (\omega-\gamma)|L_{21}|^2 + \omega|L_{31}|^2],$$
$$|L_{13}|^2 = |L_{23}|^2, \quad |L_{31}|^2 = |L_{32}|^2, \quad |L_{11}|^2 = |L_{22}|^2, \tag{9.10}$$
$$|L_{21}|^2 = |L_{12}|^2,$$
$$(\omega+2\gamma)L_{11}\bar{L}_{13} + (\omega-2\gamma)\bar{L}_{12}L_{13} + \omega L_{31}L_{33} = 0.$$

Let us rewrite the last relation, taking into account the equality $\bar{L}_{12} = L_{21}$ (see Eq. (9.6)) and indicating the arguments of the functions L_{ij} in explicit form,

$$(\omega + 2\gamma)L_{11}(\chi_1)\bar{L}_{13}(c) + (\omega - 2\gamma)L_{21}(\chi_1)L_{13}(c)$$
$$+ \omega L_{31}(\chi_1)L_{33}(c) = 0. \tag{9.11}$$

This equality is satisfied in several ways. Consider them sequentially.

9.2.1. *Flows with straight vortex lines*

Let us first highlight the case where the elements of the matrix $\hat{L}$ do not depend on c.

(a) **Vortex lines precess relative to the rotation axis**: Let $L_{33} = 1$ and consider L_{13} real, assuming it to be equal to

$$L_{13} = -\delta\omega,$$

where δ is a real constant. Then, from Eq. (9.11), the relation

$$L_{31} = \delta[(\omega + 2\gamma)L_{11} + (\omega - 2\gamma)L_{21}]$$

follows. It is seen that two of the three functions L_{11}, L_{21}, and L_{31} can be chosen arbitrarily, and the third will be their linear combination. Hereafter, for simplicity, we consider the frequencies ω and γ to be dimensionless.

Assume $L_{11} = 1$ and $L_{21} = F'$. In this case, the matrix $\hat{L}_0$ can be written as

$$\hat{L}_0 = \begin{pmatrix} 1 & \bar{F}'(\chi_2) & -\delta\omega \\ F'(\chi_1) & 1 & -\delta\omega \\ \delta[\omega + 2\gamma + (\omega - 2\gamma)F'] & \delta[\omega + 2\gamma + (\omega - 2\gamma)\bar{F}'] & 1 \end{pmatrix}.$$

Here, the functions in the first column depend on χ_1 and in the second column on χ_2. An expression of the form $\bar{F}(\chi_i)$ should be understood as a designation of the function $\overline{F(\bar{\chi}_i)}$.

Expressions for the particle trajectories corresponding to this form of $\hat{L}_0$ are given by

$$W_1 = \frac{X + iY}{\sqrt{2}} = \{\chi_1 e^{-i\omega t} + \bar{F}(\chi_2)e^{i\omega t} - \delta\omega c\}e^{-i\gamma t}, \quad W_2 = \bar{W}_1,$$

$$Z = c + 2\delta\mathrm{Re}\{[(\omega + 2\gamma)\chi_1 + (\omega - 2\gamma)F(\chi_1)]e^{-i\omega t}\}. \tag{9.12}$$

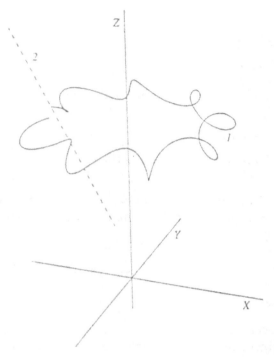

Fig. 9.1. Curve 1 is an example of particle trajectory and straight 2 is the instantaneous position of the corresponding vortex line.

The trajectories of the particles are four-frequency in time. They are windings on toroidal surfaces formed by the rotation of ellipses differently oriented in space around the z-axis (see Fig. 9.1). For incommensurable ω and γ, these windings are quasi-periodic, and each trajectory fills the entire surface of the corresponding torus. The vorticity of the flow (9.12) is found using Eq. (1.39) and has the following form:

$$\vec{\Omega} = \frac{S_3}{\det \hat{L}_0} \frac{\partial \vec{R}}{\partial c} = \frac{S_3}{\det \hat{L}_0} \begin{pmatrix} -\sqrt{2}\delta\omega\cos\gamma t \\ -\sqrt{2}\delta\omega\sin\gamma t \\ 1 \end{pmatrix}; \qquad (9.13)$$

$S_3 = -2\{\omega + \gamma + (\omega - \gamma)|F'|^2 + 4\delta^2\omega[\omega + 2\gamma + (\omega - 2\gamma)|F'|^2]\};$
$\det \hat{L}_0 = (1 + 2\delta^2\omega^2)(1 - |F'|^2) + 4\omega\gamma\delta^2|1 - F'|^2.$

The vortex lines of such a flow are straight lines precessing around the z-axis. They are "frozen" into the flow and are specified parametrically by Eq. (9.13), where the coordinate c serves as the parameter (see Fig. 9.1). At $\delta = 0$, the three-dimensional flow (9.13) is transformed into a flat Ptolemaic one (Section 7.2), in which the vorticity is directed along the z-axis.

In the theory of vortex flows, Eq. (9.12) acts as a unique example of an exact solution describing a class of spatial non-stationary fluid motion, which depends on an arbitrary analytical function and three actual parameters. According to the "measure of complexity," the solution is apparently difficult to compare with any other known exact analytical solution of fluid dynamics. At the same time, the problem of physical interpretation is extremely important for this solution. The fact is that Eq. (9.12) cannot determine the motion of a fluid in an unbounded volume since in the space of Lagrangian coordinates there will always be a point where the determinant of the matrix $\hat{L}$ becomes zero. In addition, these solutions, unlike Ptolemaic ones, cannot be linked to an external potential flow since vortex lines, which are straight lines in this case, cannot end inside the fluid (the case of bounded volume and linear function F corresponding to the flow inside an ellipsoid (see Section 3.6) is not considered here). Thus, there is a problem with the physical interpretation of solution (9.12).

We can only indicate one possible solution. Let us extract some volume in the space of Lagrangian coordinates, e.g., a unit sphere. Its shape in physical space will depend on the shape of F. If, in addition to this, we assume that the actual parameters ω and γ (or their combinations included in Eq. (9.12)) are irrational, then the shape of the volume under consideration will never be repeated. Thus, we have an exact solution that determines the dynamics of a bounded volume of fluid with a boundary shape variation law that is close to random. Based on this, it can be hypothesized that expression (9.12) can determine the motion of a fluid inside coherent structures observed in turbulent flows. And since a characteristic property of motion (9.12) is that the vortex lines are unidirectional inside the chosen fluid, it should be natural to call such localized structures *vortex domains*.

(b) **Vortex lines precess in the horizontal plane:** Assume $L_{33} = 0$. Then, the function $L_{31}(\chi_1)$ is arbitrary, and the elements

174   *Analytical Fluid Dynamics in Lagrangian Variables*

L_{11} and L_{21} of the matrix are connected by the relation

$$(\omega + 2\gamma)L_{11} + (\omega - 2\gamma)L_{21} = 0.$$

Denoting

$$L_{11} = 2\gamma - \omega, \quad L_{21} = 2\gamma + \omega, \quad L_{13} = 1, \quad \text{and} \quad L_{31} = G'(\chi_1),$$

we write the matrix $\hat{L}_0$ in the following form:

$$\hat{L}_0 = \begin{pmatrix} 2\gamma - \omega & 2\gamma + \omega & 1 \\ 2\gamma + \omega & 2\gamma - \omega & 1 \\ G' & \bar{G}' & 0 \end{pmatrix}.$$

The desired flow is described by the expressions

$$W_1 = [(2\gamma - \omega)\chi_1 e^{-i\omega t} + (2\gamma + \omega)\bar{\chi}_1 e^{i\omega t} + c]e^{-i\gamma t}.$$
$$Z = 2\mathrm{Re}[G(\chi_1)e^{-i\omega t}].$$

For this family of motions, the vorticity vector is

$$\vec{\Omega} = \frac{2\omega^2 + \gamma^2 |G'|^2}{2\gamma \mathrm{Re}\bar{G}'} \begin{pmatrix} -\sqrt{2}\cos\gamma t \\ \sqrt{2}\sin\gamma t \\ 0 \end{pmatrix}.$$

As can be seen, the vortex lines are horizontal straight lines which precess around the vertical axis.

9.2.2. *Flows with vortex lines in the same plane with the rotation axis*

Consider the case $L_{31} = 0$. For simplicity, we assume that L_{13} is real. It follows from Eq. (9.11) that

$$L_{21} = \frac{2\gamma + \omega}{2\gamma - \omega}L_{11} \quad \text{and} \quad \omega^2 \neq 4\gamma^2.$$

Denoting

$$L_{11} = (2\gamma - \omega)F'(\chi_1), \quad L_{13} = h'(c), \quad \text{and} \quad L_{33} = q'(c),$$

Spatial Vortex Flows 175

we write the matrix $\hat{L}_0$ in the form

$$\hat{L}_0 = \begin{pmatrix} (2\gamma - \omega)F'(\chi_1) & (2\gamma + \omega)F'(\chi_2) & h'(c) \\ (2\gamma + \omega)F'(\chi_1) & (2\gamma - \omega)F'(\chi_2) & h'(c) \\ 0 & 0 & q'(c) \end{pmatrix},$$

and by simple integration, we find an explicit representation for the solution

$$W_1 = \frac{X + iY}{\sqrt{2}} = [(2\gamma - \omega)F'(\chi_1)e^{-i\omega t}$$

$$+ (2\gamma + \omega)F(\chi_2)e^{i\omega t} + h(c)]e^{-i\gamma t};$$

$$W_2 = \bar{W}_1, \quad Z = q(c).$$

Note that $\chi_2 = \bar{\chi}_1$. Due to the arbitrary choice of the domain of definition of Lagrangian coordinates, it can be assumed without loss of generality that

$$F(\chi_1) = \chi_1 \quad \text{and} \quad q(c) = c.$$

Thus, finally, the solutions of the type considered have the form

$$W_1 = [(2\gamma - \omega)\chi_1 e^{-i\omega t} + (2\gamma + \omega)\bar{\chi}_1 e^{i\omega t} + h(c)]e^{-i\gamma t}, \quad Z = c. \tag{9.14}$$

The resulting motion at $h(c) = 0$ reduces to a two-dimensional Ptolemaic one (7.4). The motion of particles in flow (9.14) occurs in the $z = $ const, plane along trajectories representing the superposition of three circular rotations with different frequencies, $\omega + \gamma, \omega - \gamma$, and γ, and different amplitudes, or motion along an ellipse, the center of which moves along a circle (Fig. 9.2). In the case of incommensurable frequencies ω and γ, the trajectories are quasi-periodic and unclosed. Despite the fact that they all lie in parallel planes, the flow is three-dimensional since the coordinate of the center around which rotation occurs in each of the z-planes is determined by the function $h(c)$, i.e., it changes when moving from one layer to another. The arbitrariness of the function $h(c)$ is limited by the requirement of its unambiguity.

For this type of flow,

$$\det \hat{L}_0 = -8\gamma\omega;$$

therefore, neither γ nor ω can be assumed to be equal to zero. A special case of the considered flows is the Kirchhoff vortex. It corresponds to the case $\omega = \gamma$ and $h = 0$. In a more general situation

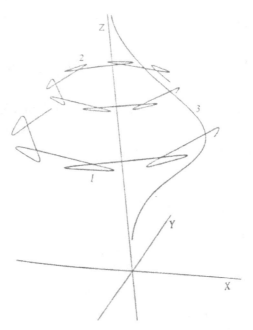

Fig. 9.2. Curves 1 and 2 are the examples of particle trajectories; 3 is the instantaneous position of one of the vortex lines.

($\omega = \gamma$ and $h \neq 0$), the flow becomes three-dimensional. Herein, the vortex axis bends in space in accordance with the shape of $h(c)$, but the vortex core itself does not deform and remains elliptical. It should be natural to call such a vortex a curvilinear Kirchhoff one. The case $\omega \neq \gamma$ corresponds to more complex rotation regimes of the core.

The vorticity of the flow (9.14) is determined by the formula

$$\vec{\Omega} = \frac{\omega^2}{2\gamma} \begin{pmatrix} \sqrt{2}h'(c)\cos\gamma t \\ \sqrt{2}h'(c)\sin\gamma t \\ 1 \end{pmatrix}. \tag{9.15}$$

The vorticity vector at any point of the flow precesses around the z-axis with frequency γ. The shape of the vortex lines is determined by Eqs. (9.14) and 9.15, where the complex variables are fixed, and c is a real parameter. In other words, the vortex lines coincide with the coordinate lines of c. They lie in the same plane with the rotation axis (e.g., at $t = 0$, the vector $\vec{\Omega}$ belongs to the xz plane).

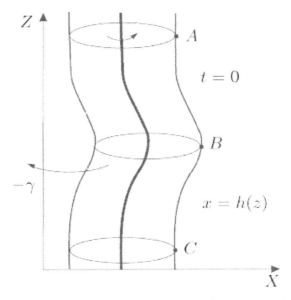

Fig. 9.3. Flow diagram in a vortex cord of rotating fluid.

In Eulerian form, flow (9.14) can be written as

$$W_t = \frac{i\omega^2}{4\gamma}W + \frac{i(\omega^2 - 4\gamma^2)}{4\gamma}\bar{W}e^{-2i\gamma t} - \frac{i\omega^2}{2\gamma}h(z)e^{-i\gamma t}; \quad Z_t = 0. \quad (9.16)$$

The first term in the expression for the complex velocity describes the circular rotation that generates the z component of vorticity, and the second corresponds to the non-stationary potential motion. The third term characterizes the difference in flows at different horizontal levels (it generates oscillating vorticity components lying in the xy plane). The Eulerian representation for vorticity can be obtained from Eq. (9.16) by replacing c with z. The vertical axis in the space of Lagrangian coordinates ($\chi_1 = \bar{\chi}_1 = 0$) forms a flat line $x = h(z)$ at $t = 0$. At the next moments of time, this line uniformly rotates around the z-axis with angular velocity $-\gamma$. Its equation at any arbitrary time is written as

$$x + iy = h(z)e^{-i\gamma t}.$$

It coincides with the vortex line and serves as a rotation axis for the flow (9.14). Figure 9.3 shows qualitatively how much volume is occupied by fluid particles located in the Lagrangian coordinate system inside a cylinder with a circular cross section. In physical space, the cross section of such a curved column is already elliptical. Its generatrix ABC at the next moments of time will remain a flat curve parallel to the axis of the column (the bold line in Fig. 9.3), i.e., the fluid particles that make up the ABC line will not swirl relative to each other. We point out that the particles inside the column can also rotate in the direction opposite to the precession of the column itself around the z-axis, which is shown in Fig. 9.3.

For a bounded $h(c)$ at sufficiently far distances from the z-axis ($|W| \gg h$), the Eulerian velocity field (9.16) will approximately have the form

$$W_t = \frac{i\omega^2}{4\gamma}W + \frac{i(\omega^2 - 4\gamma^2)}{4\gamma}\bar{W}e^{-2i\gamma t}. \tag{9.17}$$

It represents the sum of a term that describes uniform rotation with frequency $\omega^2/4\gamma$ and a deformation field that is linear in coordinates and harmonic in time. Thus, flow (9.14) for a bounded $h(c)$ can be interpreted as a possible regime of fluid motion in an external quasi-two-dimensional field with asymptotic form (9.17). Due to the constraint on the choice of constants ω and γ, flow (9.16) cannot be purely circular, but for ω values that differ only slightly from $\pm 2\gamma$, it can be arbitrarily close to uniform rotation at a great distance.

9.2.3. *Flows for which the rotation axis is inclined to the vortex line plane*

Let the function L_{13} be equal to

$$L_{13} = A'(c) + iB'(c),$$

where A and B are the single-valued functions, $L_{33} = 1$, and all other elements of the matrix $\hat{L}_0$ are constant values. From Eq. (9.11), under the condition $L_{31} \neq 0$, it follows that

$$\frac{(\omega + 2\gamma)L_{11} + (\omega - 2\gamma)L_{21}}{\omega L_{31}}A'$$

$$+ i\frac{(\omega - 2\gamma)L_{21} - (\omega + 2\gamma)L_{11}}{\omega L_{31}}B' = -1. \tag{9.18}$$

Assume that

$$\text{Im}\frac{(\omega + 2\gamma)L_{11} + (\omega - 2\gamma)L_{21}}{\omega L_{31}}$$
$$= \text{Re}\frac{(\omega - 2\gamma)L_{21} - (\omega + 2\gamma)L_{11}}{\omega L_{31}} = 0. \tag{9.19}$$

Then, one of the functions, e.g., $A(c)$, can be considered arbitrary, while the other will be determined from Eq. (9.18). Let $L_{11} = 1$. Taking into account Eqs. (9.18) and (9.19), we find the relation between A and B:

$$B = \frac{\text{Re}\,\varepsilon}{\text{Im}\,\varepsilon}A + \frac{|\varepsilon|^2}{2\text{Im}\,\varepsilon(1 + 2\gamma/\omega)}c; \quad L_{31} = \varepsilon.$$

The general form of the matrix $\hat{L}_0$ for this type of solution is now written as

$$\hat{L}_0 = \begin{pmatrix} 1 & \dfrac{\bar{\varepsilon}\left(1 + \frac{2\gamma}{\omega}\right)}{\varepsilon\left(1 - \frac{2\gamma}{\omega}\right)} & A'\left[1 + i\dfrac{\text{Re}\,\varepsilon}{\text{Im}\,\varepsilon}\right] + \dfrac{i|\varepsilon|^2}{2\text{Im}\,\varepsilon\left(1 + \frac{2\gamma}{\omega}\right)} \\[4mm] \dfrac{\varepsilon\left(1 + \frac{2\gamma}{\omega}\right)}{\bar{\varepsilon}\left(1 - \frac{2\gamma}{\omega}\right)} & 1 & A'\left[1 - i\dfrac{\text{Re}\,\varepsilon}{\text{Im}\,\varepsilon}\right] - \dfrac{i|\varepsilon|^2}{2\text{Im}\,\varepsilon\left(1 + \frac{2\gamma}{\omega}\right)} \\[4mm] \varepsilon & \bar{\varepsilon} & 1 \end{pmatrix}. \tag{9.20}$$

The solution of the fluid dynamics equations, which corresponds to the constructed matrix, for an arbitrary moment of time is obtained by substituting $\hat{L}_0$ into formula (9.8). The determination of the trajectories of fluid particles is carried out by integrating over complex Lagrangian coordinates and gives the following result:

$$W = \left[\chi_1 e^{-i\omega t} + \frac{\bar{\varepsilon}\left(1 + \frac{2\gamma}{\omega}\right)}{\varepsilon\left(1 - \frac{2\gamma}{\omega}\right)}\chi_2 e^{i\omega t} + \left(1 + i\frac{\text{Re}\,\varepsilon}{\text{Im}\,\varepsilon}\right)A(c)\right.$$
$$\left. + \frac{i|\varepsilon|^2 c}{2\text{Im}\,\varepsilon\left(1 + \frac{2\gamma}{\omega}\right)}\right] e^{-i\gamma t}, \tag{9.21}$$

$$Z = \varepsilon\chi_1 e^{-i\omega t} + \bar{\varepsilon}\bar{\chi}_1 e^{i\omega t} + c; \quad \varepsilon \neq 0.$$

180   *Analytical Fluid Dynamics in Lagrangian Variables*

The trajectories of fluid particles represent spatial curves, which depend on four time frequencies. The motion of an individual particle can be represented as the superposition of three rotations in the horizontal plane with frequencies γ, $\omega + \gamma$, and $\omega - \gamma$ and rectilinear oscillations in the vertical plane.

The determinant of the matrix $\hat{L}_0$ is equal to

$$\det \hat{L}_0 = 4\omega\gamma \frac{(2\gamma - \omega)|\varepsilon|^2 - 2(\omega + 2\gamma)}{(\omega - 2\gamma)^2(\omega + 2\gamma)}.$$

It does not depend on the function $A(c)$ and is equal to a constant value. This means that solution (9.21) exists in the entire space. To find the vorticity, we determine the only non-zero Cauchy invariant (see Eq. (9.13))

$$S_3 = -\frac{2\omega[2\omega^2 + (\omega - 2\gamma)^2|\varepsilon|^2]}{(\omega - 2\gamma)^2}$$

and use Eq. (1.38) for vorticity

$$\vec{\Omega} = \frac{S_3}{\det \hat{L}_0} = \frac{(\omega + 2\gamma)[2\omega^2 + (\omega - 2\gamma)^2|\varepsilon|^2]}{2\delta[(\omega - 2\gamma)|\varepsilon|^2 + 2(\omega + 2\gamma)]}\vec{R}_c = \beta\vec{R}_c,$$

where the vector on the right-hand side has the components (see Eq. (9.21))

$$\left\{ \begin{array}{c} A'\cos\gamma t + \left(A'\dfrac{\mathrm{Re}\varepsilon}{\mathrm{Im}\varepsilon} + \dfrac{\omega|\varepsilon|^2}{2\mathrm{Im}\varepsilon(\omega + 2\gamma)}\right)\sin\gamma t \\[3mm] -A'\sin\gamma t + \left(A'\dfrac{\mathrm{Re}\varepsilon}{\mathrm{Im}\varepsilon} + \dfrac{\omega|\varepsilon|^2}{2\mathrm{Im}\varepsilon(\omega + 2\gamma)}\right)\cos\gamma t \\[3mm] 1 \end{array} \right\}.$$

The equation of vortex lines in parametric form will have the form

$$X(c) = A\beta\cos\gamma t + \left(A\frac{\mathrm{Re}\varepsilon}{\mathrm{Im}\varepsilon} + \frac{\omega|\varepsilon|^2 c}{2\mathrm{Im}\varepsilon(\omega + 2\gamma)}\right)\sin\gamma t,$$

$$Y(c) = -A\beta\sin\gamma t + \left(A\frac{\mathrm{Re}\varepsilon}{\mathrm{Im}\varepsilon} + \frac{\omega|\varepsilon|^2 c}{2\mathrm{Im}\varepsilon(\omega + 2\gamma)}\right)\cos\gamma t, \qquad (9.22)$$

$$Z(c) = \beta c.$$

Spatial Vortex Flows 181

The torsion of the curve (9.22) is proportional to the mixed product of the vectors $\vec{R}_c, \vec{R}_{cc}$, and $\vec{R}_{ccc}$, which in this case is zero:

$$(\vec{R}_c, \vec{R}_{cc}, \vec{R}_{ccc}) = \begin{vmatrix} X_c & Y_c & 1 \\ X_{cc} & Y_{cc} & 0 \\ X_{ccc} & Y_{ccc} & 0 \end{vmatrix} = 0.$$

This is a necessary and sufficient condition for the curve to be flat. Consequently, the vortex line (9.22) is flat. However, the lines no longer lie in the same plane as the rotation axis. The angle θ between the normal to the plane of the vortex lines

$$\vec{N} = \frac{[\vec{R}_c \times \vec{R}_{cc}]}{|[\vec{R}_c \times \vec{R}_{cc}]|}$$

and the z-axis is determined by the relation

$$\cos \theta = -\frac{\omega |\varepsilon|}{\sqrt{4(\omega + 2\gamma)^2 + \omega^2 |\varepsilon|^2}}.$$

In the case $\varepsilon = 0$, the vortex line lies in the same plane with the rotation axis. This case corresponds to the solution (9.14). The passage to the limit in Eq. (9.21) should be performed in two stages: First, we put $\mathrm{Re}\,\varepsilon = 0$ and then let $\mathrm{Im}\,\varepsilon$ go to zero.

Expression (9.21) describes a complex spatial flow with flat curvilinear vortex lines precessing around a certain axis. It generalizes solution (9.14). For given values of the frequencies ω and γ, the shape of the vortex lines in it is determined by the shape of the function A (analog of h in Eq. (9.14)), and the angle of deviation from the rotation axis is determined by the parameter ε.

The solutions obtained in Chapters 7–9 were based on a complex representation of the trajectories of fluid particles. A method for finding explicit solutions to Euler equations in real variables was proposed in other studies [14–16]. The matrix representation of the flow field and the method of separation of variables were used in those studies in a similar way. Unfortunately, the authors of those studies did not dwell on the issues of the novelty of the solutions obtained and their physical interpretation.

9.3. Ptolemaic Vortices in the Axial Flow

Let us mention one more version of the generalization of Ptolemaic flows [17]. It is related to the possibility, already mentioned in Section 3.3.4, of "transforming" plane (x, y) flows into spatial ones by adding to the solution the non-uniform drift of fluid particles along the z-axis.

Assume that the flow inside the vortex core is described by the following expressions (see Eq. (8.1)):

$$W = X + iY = \alpha\nu\exp(i\lambda t) + F(\bar{\nu}); \quad \nu = \exp[ik(a + ib)],$$
$$Z = c + U(|\nu|^2)t; \quad |\nu| \leq 1; \quad U(1) = 0. \tag{9.23}$$

The form of the axial flow was chosen in such a way that the flow velocity along the z-axis becomes zero at the boundary of the vortex. In this case, the potential flow outside the vortex will still be specified by Eq. (8.3). However, the "internal" motion of the fluid will change. Fluid particles will move along helical lines with a pitch of $2\pi\lambda^{-1}U(|\nu|^2)$.

Expressions for the vorticity of a given flow are found by substituting Eq. (9.23) into Eq. (1.31). For the components of the vorticity vector lying in the x, y plane, the following representation is valid:

$$\begin{aligned}
\Omega_x + i\Omega_y &= \left[\frac{D(W, \bar{W})}{D(\nu, \bar{\nu})}\right]^{-1} \frac{D(W, U)}{D(\nu, \bar{\nu})} \\
&= 2i\frac{U_\nu F'(\bar{\nu}) - \alpha U_{\bar{\nu}}\exp(i\lambda t)}{\alpha^2 - |F'|^2}.
\end{aligned} \tag{9.24}$$

Each of the components Ω_x and Ω_y has both oscillating with frequency λ and time-independent components. The axial component of vorticity remains the same as in the planar case and is determined by Eq. (8.2).

To determine the Cauchy invariants, it is easiest to substitute solution (9.23) into Eq. (1.23). From them, we find that

$$S_1 = U_b'; \quad S_2 = -U_a'; \quad S_3 = 2\lambda.$$

The second of these formulas takes into account that $|\nu|^2 = \exp(-2kb)$ does not depend on a. Unlike the generalized Ptolemaic flows discussed earlier in this chapter, three Cauchy invariants are already non-zero for motion (9.23).

The shape of vortex lines along the known vector $\vec{\Omega}$ is determined from the equation

$$\frac{dx}{\Omega_x} = \frac{dy}{\Omega_y} = \frac{dz}{\Omega_z}. \tag{9.25}$$

It is difficult to integrate it in the general case of flows (9.23) since it is impossible to explicitly write the dependence of the vorticity components on the Eulerian coordinates x, y, and z. As a result, we limit ourselves to analyzing the shape of vortex lines for the simplest case, namely, the Kirchhoff vortex corresponding to the linear function $F(\bar{\nu}) = \beta\bar{\nu}$. Let us choose the axial flow in the form $U(|\nu|^2) = U_0(1 - |\nu|^2)$. In a reference frame rotating with angular velocity $\lambda/2$, this flow is stationary and the vortex lines are fixed. Therefore, in the laboratory system, remaining unchanged, the vortex lines will rotate with the same frequency. To find their shape, it is enough to analyze the vorticity distribution at one of the moments of time, e.g., when $t = 0$.

Assume that $t = 0$ in Eqs. (9.23) and (9.24) and find the vorticity distribution at the initial moment depending on the Eulerian coordinates. The vorticity distribution has the following form:

$$\Omega_x = -\frac{2U_0}{(\alpha - \beta)^2}y; \quad \Omega_y = \frac{2U_0}{(\alpha + \beta)^2}x; \quad \Omega_z = \frac{2\lambda\alpha^2}{\alpha^2 - \beta^2}.$$

We now substitute these relations into Eq. (9.25):

$$-(\alpha - \beta)^2\frac{dx}{y} = (\alpha + \beta)^2\frac{dy}{x} = \frac{(\alpha^2 - \beta^2)U_0}{\lambda\alpha^2}dz;$$

its solution in parametric form will be written as

$$x = (\alpha + \beta)\cos s, \quad y = (\alpha - \beta)\sin s, \quad z = \lambda\alpha^2 s.$$

This is the equation of a helical line uniformly wound with a pitch of $2\pi\lambda\alpha^2$ on the surface of an elliptical cylinder with semi-axes $(\alpha + \beta)$ and $(\alpha - \beta)$. At times $t > 0$, the vortex line retains its shape, rotating with respect to the initial position by an angle $\lambda t/2$.

For nonlinear F, the vortex lines will still be helices, but they will already be wound on more complex surfaces, and their pitch will generally be non-uniform.

References

[1] Batchelor, G.K. (1967). *An Introduction to Fluid Dynamics* (Cambridge University Press, Cambridge).

[2] Shafranov, V.D. (1958). On magnetohydrodynamical equilibrium configurations. *J. Exper. Theor. Phys.*, 6(3), pp. 710–722.

[3] Landau, L.D. and Lifshitz, E.M. (1960). *Electrodynamics of Continuous Media* (Pergamon Press, New York).

[4] Kadomtsev, B.B. (1982). *Collective Phenomena in Plasmas* (Elsevier Science Limited).

[5] Hill, M.J.M., (1884). On the motion of fluid, part of which is moving rotationally and part irrotationally. *Phil. Trans. R. Soc. A, 175*, 363–410.

[6] Hicks, W.M. (1899). Researches in vortex motion. On spiral or gyrostatic vortex aggregates. *Phil. Trans. Roy. Soc. A*, 192, pp. 33–101.

[7] Moffatt, H.K. (1969). The degree of knottedness of tangled vortex lines. *J. Fluid Mech.*, 35, pp. 117–129.

[8] Shmyglevsky, Yu.D. (1993). On swirling flows of ideal and viscous fluids. *Zhurn. Vychisl. Matem. Mat. Fiz.*, 3312, pp. 1905–1911 [in Russian].

[9] Skvortsov, A.T. (1988). Exact solution of the magnetic hydrodynamics equations in the form of a solitary toroidal vortex. *Pisma Zhurn. Eksp. Teor. Fiz.*, 14(17), pp. 1609–1611 [in Russian].

[10] Kaptsov, O.V. (1989). Some classes of two-dimensional vortex flows of an ideal fluid. *J. Appl. Mech. Tech. Phys.*, 30(1), pp. 105–112.

[11] Andreev, V.K., Kaptsov, O.V., Pukhnachev, V.V. and Rodionov, A.A. (1998). *Application of Group-theoretic Methods in Fluid Dynamics. Mathematics and Applications,* Vol. 450 (Kluwer Academic Publishers, Dordrecht).

[12] Abrashkin, A.A., Zen'kovich, D.A. and Yakubovich, E.I. (1996). Matrix formulation of hydrodynamics and extension of Ptolemaic flows to three-dimensional flows. *Radiophys. Quant. Electronics, 39*, pp. 518–526.

[13] Abrashkin, A.A., Zen'kovich, D.A. and Yakubovich, E.I. (1997). Study of three-dimensional vortex flows using matrix hydrodynamic equations. *Phys.- Dokl.*, 42(12), pp. 687–690.

[14] Martín, M. and Tuomela, J. (2019). 2D incompressible Euler equations: New explicit solutions. *Discr. Contin. Dyn. Syst. A,* 39(8), pp. 4547–4563.

[15] Saleva, T. and Tuomela, J. (2021). On the explicit solutions of separation of variables type for the incompressible 2D Euler equations. *J. Math. Fluid Mech.*, 23, 39.

[16] Saleva, T. and Tuomela, J. (2022). Explicit solutions to the 3D incompressible Euler equations in Lagrangian formulation. *J. Math. Fluid Mech.*, 24, 98.

[17] Zen'kovich, D.A. (2002). Matrix Lagrangian description of vortex structures in an ideal fluid, *Dissertation of a candidate of physical and mathematical sciences* (Nizhny Novgorod), [in Russian].

Part IV
Generalized Gerstner Waves

Chapter 10

Waves on Water at Unstable Pressure

> "The wind blowing over the water
> surface generates waves through
> physical processes, which cannot be
> considered known."
>
> — F. Ursell [1]

In the theory of waves on water, only two exact solutions of the complete equations of dynamics of an ideal fluid are known. The first one was obtained by Gerstner and describes trochoidal gravity waves in deep water (see Chapter 5). The second one belongs to Crapper [2] and refers to the case of purely capillary steady-state waves.

The conventional formulation of the problem assumes that the pressure on a free surface is constant (dynamic boundary condition). It is valid in the absence of wind. The wind creates a variable pressure on the fluid surface. However, it is extremely difficult to describe the self-consistent interaction of wind and waves (see the epigraph). A possible way to overcome this difficulty is to choose a certain law of pressure variation on the surface. In other studies [3–8], where the dynamics of weakly nonlinear wave trains was studied, the external pressure was chosen in accordance with the linear theory of excitation of Miles waves [9]. Yen and Ma numerically studied the formation of strongly nonlinear waves within the framework of a complete system of equations of fluid dynamics under the assumption of a homogeneous air flow and phenomenologically set the law of pressure distribution on a free surface [10]. In yet another study [11], the effect of wind is taken into account by applying a linear relation between pressure and local surface steepness for those sections of the profile where it exceeds a certain threshold value. Thus, the effect of the wind is modeled by an inhomogeneous distribution of pressure on the surface. In the cycle of problems to be discussed in the following, we assume that the surface pressure is inhomogeneously distributed over the surface and depends harmonically on time. This law of pressure variation corresponds to the Ptolemaic flows.

Assume that the region of wave motion of a fluid in Lagrangian variables corresponds to the lower half-space ($b \leq 0$), and the flow is described by the following expression:

$$W = G(\chi) + F(\overline{\chi})\exp(-i\omega t), \quad \text{Im } \chi \leq 0. \tag{10.1}$$

The flow belongs to the family of Ptolemaic flows (7.4), and in the exponent, it was chosen for convenience that $\mu = -\omega$. The function $G(\chi)$ should be single valued. This is true if its derivative $G'(\chi)$ does not become zero in the flow region. This condition is automatically fulfilled if

$$D = |G'|^2 - |F'|^2 > 0. \tag{10.2}$$

Due to the fact that the case $\delta = 0$ is being considered, fluid particles now rotate relative to their equilibrium position in a circle. The rotation radius is determined by the modulus of the function F. At the depth, the particles are at rest, so the condition

$$|F| \to 0 \quad \text{at } b \to -\infty \tag{10.3}$$

should be fulfilled. Since the function F is analytical, its modulus reaches its largest value at the free boundary. Therefore, the maximum value of oscillations of a single fluid particle will correspond to the particles located on the free surface.

The wave solution (10.1) corresponds to the following pressure distribution (see Eq. (7.9)):

$$\frac{p - p_0}{\rho} = -g\text{Im}(G + Fe^{-i\omega t}) + \frac{\omega^2}{2}|F|^2 + \text{Re}\left(e^{-i\omega t}\int \omega^2 G'\overline{F}d\chi\right). \tag{10.4}$$

It can be seen from this ratio that at $b = 0$ the pressure will depend on the horizontal Lagrangian coordinate a and change periodically over time. The type of inhomogeneous distribution over the free surface is determined by the functions G and F. Equation (10.1) determines a whole class of exact solutions describing the dynamics of a free surface under inhomogeneous and harmonically varying pressure.

The nonlinear waves found do not carry out mass transfer, i.e., fluid particles move in a circle and there is no drift flow. The vorticity of the flow (10.1) is determined by the expression

$$\Omega = \frac{2\omega |F'|^2}{|G'|^2 - |F'|^2},$$

(10.5)

and, by virtue of condition (10.2), it is always sign constant. The vorticity is maximum at the free boundary and decreases with depth to zero (at the bottom).

Singular points (cusps) may appear on the free surface profile of the wave. A necessary condition for the occurrence of such a singular point is the appearance of a vertical tangent on the profile, which for this solution corresponds to the vanishing of the Jacobian D, which is possible only on the free surface at points satisfying the relation $|G'(a)| = |F'(a)|$. In this case, the Jacobian D can become zero only at discrete times.

10.1. Oscillating Standing Soliton

The case of a linear function G and an exponential function F (Gerstner wave) corresponds to the simplest choice. This is the only case where the solution (10.1) describes a stationary wave on the water. The arbitrariness in the choice of function F (for linear G) permits one to consider a wide range of possible initial shapes of the free surface profile.

Consider the Ptolemaic solution of the following form:

$$W = \chi + \frac{\beta}{(\overline{\chi} + i)^n} e^{-i\omega t}; \quad \beta > 0; \quad n \geq 2.$$

(10.6)

It describes the dynamics of a single peak. The function F in this case has a pole of the order n. It corresponds to the value $b = 1$, and therefore lies outside the flow region. In this formula, the quantities χ and β are chosen dimensionless. According to Eq. (10.2), the range of acceptable parameters is described by the condition $\beta \leq 1/n$.

The pressure on the free surface for the flow is written as follows:

$$\frac{p-p_0}{\rho} = -g\text{Im}\frac{\beta\exp(-i\omega t)}{(a+i)^n} + \frac{\beta\omega^2}{2(a^2+1)^n} - \frac{\beta\omega^2}{n-1}\text{Re}\frac{\exp(i\omega t)}{(a-i)^{n-1}}. \quad (10.7)$$

It decreases with increasing distance from the soliton and is equal to a constant value at infinity. In Eq. (10.7), the quantity $n \geq 2$. If we assume $n = 1$, the pressure at infinity will increase logarithmically. This solution has no physical meaning.

Figure 10.1 shows a picture of the evolution of the free surface and a diagram of the pressure variation as a function of time for the parameter values $n = 2$ and $\beta = 0.4$. In this figure, as in all the following ones, the bold lines correspond to the profile of the free surface of the wave and the lighter ones indicate the pressure distribution over the surface. The pressure curve is constructed for the normalized pressure $p_n = p/(\rho g L)$, where L is the unit length scale. For ease of analysis, the curve is shifted downward to a constant level.

At time $t = \pi/(2\omega)$, the free surface has the form of a single spike with small dips on the sides. After a quarter of the period, a symmetric over-oscillation forms on the surface, and its maximum, compared with the initial soliton profile, shifts to the right and decreases in amplitude. After another quarter of the period, a single dip with two small lateral spikes appears in the profile, and finally, after passing through the full period of oscillation, the profile returns to its original shape. The pressure on the surface varies in antiphase with the profile of the free surface. Such a non-stationary structure can be considered an oscillating standing soliton.

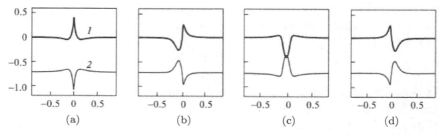

Fig. 10.1. Profile (1) and pressure (2) for a single peak without cusp: $F = 0.4/(\overline{\chi}+i)^2; (a-d): \omega t = \pi/2, \pi, 3\pi/2, 2\pi$.

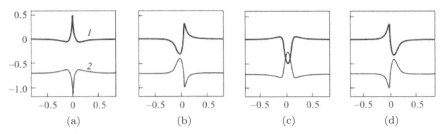

Fig. 10.2. Profile (1) and pressure (2) for a single peak without cusp: $F = 0.5/(\overline{\chi} + i)^2$; $(a - d) : \omega t = \pi/2, \pi, 3\pi/2, 2\pi$.

Figure 10.2 corresponds to the case $\beta = 0.5$. This is the maximum permissible value of parameter β. The parametric representation of the shape of the free surface in this case has the following form:

$$X_s = a + \frac{1}{2(a^2 + 1)}\left(\cos\omega t - \frac{2a}{a^2 + 1}\sin\omega t\right),$$
$$Y_s = -\frac{1}{2(a^2 + 1)}\left(\sin\omega t + \frac{2a}{a^2 + 1}\cos\omega t\right). \tag{10.8}$$

On the profile of the free surface at times $t = (\pi/2 + 2\pi l)/\omega$, where l is an integer, cusps are formed at the point corresponding to the Lagrangian coordinate $a = 0$. A cusp also appears on the pressure profile at this point at this moment. In antiphase, at $t = (3\pi/2 + 2\pi l)/\omega$, where the spike transforms to a dip, neither the profile nor the pressure distribution has a cusp.

The evolution of the free surface shape in the example under consideration is in good agreement with the breather profile obtained using numerical calculations within the framework of the Euler equation [12]. At times $t = 2\pi l/\omega$, the oscillating soliton has a smooth dip and a sharper spike, and it turns into a symmetric profile with two dips and a sharper ridge as time proceeds. We emphasize that, unlike other work [12], the solution (10.6) is vortex. The vorticity for it is determined by the equality

$$\Omega = \frac{2\omega n^2 \beta^2}{[a^2 + (1-b)^2]^{n+1} - n^2\beta^2}.$$

In contrast to Gerstner waves, where vorticity is expressed in terms of exponentially decreasing functions, the vorticity of the soliton

194   *Analytical Fluid Dynamics in Lagrangian Variables*

depends on the coordinates polynomially and decreases rather slowly when moving away from it.

10.2. Soliton Against the Background of Uniform Undulation

In solution (10.1), with the linear function G chosen, the function F can be chosen sufficiently arbitrarily. In a sense, the principle of superposition is valid for it, i.e., if we choose F as the sum of two functions, the profile of the free surface will qualitatively correspond to the superposition of the profiles determined by individual terms. Let us choose the Ptolemaic flow in the following form:

$$W = \chi + \left[iA\exp(ik\overline{\chi}) + \frac{\beta}{(\overline{\chi} + i)^n} \right] \exp(-i\omega t). \tag{10.9}$$

The function F combines the terms describing the Gerstner wave and the oscillating soliton (the quantity k is considered dimensionless here). Equation (10.9) describes a single peak against the background of stationary undulation. A simple analytical estimate for the range of acceptable values of the parameter β leads to the inequality

$$\beta \leq \frac{1 - kA}{n}. \tag{10.10}$$

More exact constraints on the β value can be obtained by numerical simulation. Figure 10.3 shows the boundaries of the parameter region for the β values as functions of kA for the cases $k = 1, 2, 3$ and $n = 2$. All of them are convex arcs based on a segment determined by Eq. (4.2). The parameters corresponding to these curves give a solution with a profile sharpening at one of the time points (once per oscillation period).

Figure 10.4 shows the profile of the free surface and pressure distribution for the case $A = 0.04, k = 1, \beta = 0.48$, and $n = 2$. With the parameter values chosen, the peak at the moment $t = \pi/2\omega$ is 12 times higher than the level of uniform undulation. No sharpening on the ridge will occur. In general, the evolution of such a soliton is similar to the behavior of an oscillating soliton in the absence of uniform undulation (Figs. 10.1 and 10.2). Away from the soliton, the pressure perturbation tends to zero as $1/a$. It is mainly concentrated in the

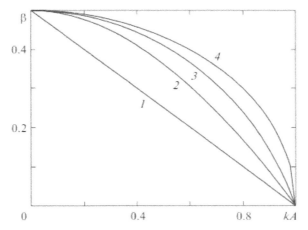

Fig. 10.3. The boundary of the range of acceptable parameters for a single peak against the background of uniform undulation; (1) $\beta = (1 + kA)/2$; (2-4) $k = 1, 2, 3$.

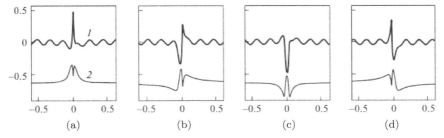

Fig. 10.4. Profile (1) and pressure (2) for a single peak against the background of uniform undulation: $A = 0.04; \beta = 0.48; n = 2; (a - d)\omega t = \pi/2, \pi, 3\pi/2,$ and 2π.

soliton region. The pressure varies in antiphase with the variation in the profile height.

The pictures showing the dynamics of a single peak against the background of the Gerstner wave can be interpreted as a variant of the formation and decay of a wave of anomalously high amplitude. The region of a sharp spike corresponds to the local minimum pressure. The law of pressure variation over a full period in the soliton region is quite complex, but the existence of a solution of form (10.9) proves the fundamental possibility of the formation of extreme waves

196 — Analytical Fluid Dynamics in Lagrangian Variables

under the action of non-uniformly distributed pressure on the surface.

As in the problem with a soliton without undulation (Figs. 10.1 and 10.2), there is a dip in front of the peak, which increases during half the period. A single spike of sufficiently high amplitude with a dip in front of the wave corresponds to the description of the so-called "Lavrenov wave" [13]. The dip enhances the destructive effect of the wave since the vessel attacked by it buries its nose into the dip and then is hit by a wave of anomalous height.

By adding a finite number of terms having a pole outside the flow region to the function F, it is possible to construct a solution for a wave train with any predetermined number of maxima. At the same time, it is obvious that the nature of the pressure variation on the surface will be more difficult the more such terms are taken into account.

10.3. Non-Stationary Gerstner Waves

Let us choose the Ptolemaic flow in the following form:

$$W = G(\chi) + iA \exp i(k\bar{\chi} - \omega t) \quad \text{and} \quad \text{Im } \chi = b \le 0. \qquad (10.11)$$

This expression is a special case of Eq. (10.1). The shape of the function F is the same as that of the Gerstner wave, and the function G may vary. If it is linear, then Eq. (10.11) passes into the classical Gerstner solution (see Chapter 5).

The flow (10.11) has a number of properties in common with the Gerstner wave:

— fluid particles move along circles of radius $A \exp(kb)$;
— on each Lagrangian horizon $\text{Im } \chi = b = \text{const.}$, fluid particles rotate in circles of the same radius;
— there is no averaged drift of fluid particles;
— the waves have vorticity.

However, there are some differences. For flow (10.11), the vorticity is written as (see Eq. (10.5))

$$\Omega = \frac{2\omega(kA)^2 \exp(2kb)}{|G'|^2 - (kA)^2 \exp(2kb)}. \qquad (10.12)$$

For the Gerstner wave, the vorticity depends only on the b coordinate, whereas Eq. (10.12) is a function of both Lagrangian coordinates.

The vertical coordinate of the centers of rotation of fluid particles $Y_c = \text{Im } G(a + ib)$ depends on the shape of G and also depends on both Lagrangian variables. For particles of the free surface, the centers of rotation lie on the line $Y_c^* = \text{Im } G(a)$, which is no longer horizontal, as in the case of the Gerstner wave. Thus, the particles of the free surface rotate along circles of the same radius, the centers of which lie at different levels. As a result, the wave profile changes, so it is natural to call the waves (10.11) non-stationary Gerstner waves.

Here is an expression for pressure:

$$\frac{p - p_0}{\rho g} = \frac{\omega^2}{2g} A^2 e^{2kb} + \text{Im} \left[A e^{i\omega t} \left(\frac{\omega^2}{g} \int G' e^{-ik\chi} d\chi - i e^{-ik\chi} \right) - G \right].$$

In the Gerstner wave, $G(\chi) = \chi, \omega^2 = gk$, so that the expression in the parentheses is zero, and the pressure depends only on the vertical Lagrangian coordinate b and is constant on the free surface ($b = 0$). For a non-stationary Gerstner wave, the pressure on the free surface changes harmonically with time. We attribute this to the action of the wind. Since the function G is arbitrary, the pressure distribution along the profile can be specified as arbitrarily complex.

However, there are still constraints on the choice of G. This is the condition (10.2), which in this case has the form

$$|G'| \geq kA. \tag{10.13}$$

The equality sign (a singularity of the vorticity field, see Eq. (10.12)) corresponds to the formation of cusps on the profile. Inequality (10.13) slightly reduces the possibilities of specifying the function G, and, obviously, Eq. (10.11) determines a family of waves, the shape of which depends on an arbitrary analytical function to a large extent.

We now focus on one of its particular representations. Let the variable pressure act on a limited area of the free surface. This requirement corresponds to the following asymptotic behavior:

$$G(\chi) \to \chi \quad \text{at } \chi \to \pm\infty.$$

At both infinities, the wave under study will asymptotically tend to the Gerstner wave. Since the pressure is constant in these regions, we can assume that the dispersion relation of our wave is the same as that of the Gerstner wave.

Let us choose the function G in the form

$$G(\chi) = \chi + \frac{\beta}{\chi - i\alpha}; \quad \alpha, \beta = \text{const} > 0, \quad \text{Im } \chi \leq 0. \qquad (10.14)$$

The parameters α and β have dimensions L and L^2, respectively. The function G has a pole at the point $\chi = i\alpha$, but it is outside the flow region. The derivative G' becomes zero at the points $\chi_\pm = \pm\sqrt{\beta} + i\alpha$, which also lie outside this region. Thus, the function G is bounded and single valued in the Lagrangian half-plane $\text{Im } \chi \leq 0$.

Consider now the sign constancy condition of the Jacobian D. The function G' reaches an extreme value on the free boundary, where the square of its modulus is expressed as

$$|G'(a, b = 0)|^2 = 1 - \frac{2\beta}{a^2 + \alpha^2} + \frac{\beta^2 + 4\beta\alpha^2}{(a^2 + \alpha^2)^2}.$$

The right-hand side of this ratio has an extremum at points satisfying the condition $a_*^2 = \beta + 3\alpha^2$. Its value is $(4\alpha^2/\beta + 4\alpha^2) < 1$. Since in the middle and at the ends of the variation interval of a the relations

$$|G'(a = 0, b = 0)|^2 = \left(1 + \frac{\beta}{\alpha}\right)^2 > 1$$

and

$$|G'(a = \pm\infty, b = 0)|^2 = 1$$

are fulfilled, the points $\pm a_*, b = 0$ act as the coordinates of the minimum of the function G' and the condition (10.13) can be written in the form

$$\beta \leq \frac{4(1 - k^2 A^2)\alpha^2}{k^2 A^2}.$$

The equality sign corresponds to the wave of extreme steepness:

$$kA = \frac{2\alpha}{\sqrt{\beta + 4\alpha^2}}. \qquad (10.15)$$

In contrast to the stationary Gerstner wave, this parameter is always less than unity. It also follows from Eq. (10.15) that in the intervals

Waves on Water at Unstable Pressure 199

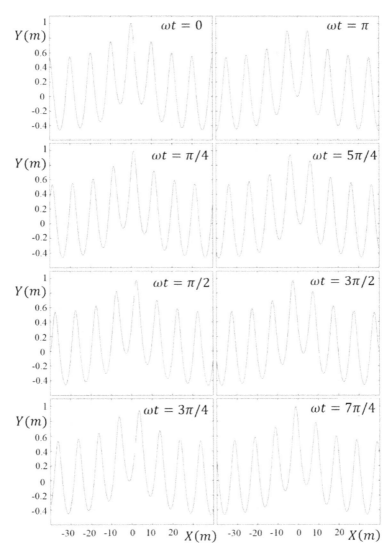

Fig. 10.5. Profile of the wave at different times. The zero of the Y coordinate corresponds to the average level of a homogeneous wave (measured in meters). The line $X = 0$ corresponds to the axis of symmetry of the wave at the initial moment.

where the wave train is homogeneous, the wave profile has no sharp points.

The dynamics of the free surface during the period is shown in Fig. 10.4. The wave profile $Y(X)$ is parametrically specified as

$$X = a + \frac{\beta a}{a^2 + \alpha^2} - A\sin(ka - \omega t);$$

$$Y = \frac{\beta \alpha}{a^2 + \alpha^2} + A\cos(ka - \omega t).$$

The following parameter values were specified: $A = 0.5\,m$, $\lambda = 10\,m$, $\alpha = 10\,m$, and $\beta = 5\,m^2$. The curves in Fig. 10.5 correspond to the change of ωt by $\pi/4$. The wave profile has the form of an inhomogeneous wave train, which is symmetric at the initial moment of time with the vertical coordinate at the central maximum $(X = 0)$, equal to 1 m, being twice the amplitude of the homogeneous train A. The width of the disturbed part of the train, where the amplitudes of the maxima exceed this value, is of the order of 8λ. Profile variations inside this part are clearly seen in the figure. The amplitude of the maximum located behind the central maximum increases, and the latter, in turn, decreases its peak. After half a period, the train restores its symmetric shape, but now we have a dip at the point $= 0$. The adjacent maxima have the same height of about $0.9\,m$. Subsequently, the dip starts to shift to the right, and the maximum behind it increases. At $\omega t = 2\pi$, the wave takes its original shape.

In the region of an inhomogeneous train, where the height of the maxima exceeds $0.5\,m$, the average fluid level lies above the level $Y = 0$. This is due to the choice of the G function. The pressure in this region differs from constant (atmospheric) pressure. It is convenient to express this quantity as

$$\Delta p_* = \frac{p - p_0}{\rho g A} = \frac{1}{2} kA - \frac{\beta \alpha}{A(a^2 + \alpha^2)} + k\beta \int_{-\infty}^{a} [(a^2 - \alpha^2)\sin(ka - \omega t)$$
$$+ 2a\alpha\cos(ka - \omega t)](a^2 + \alpha^2)^{-2} da,$$

where $p_0 = p_a - \frac{1}{2}\rho g k A^2$ and p_a is the atmospheric pressure. The dynamics of surface pressure in the region of an inhomogeneous train is shown in Fig. 10.6. Its distribution has the form of a pit, the depth

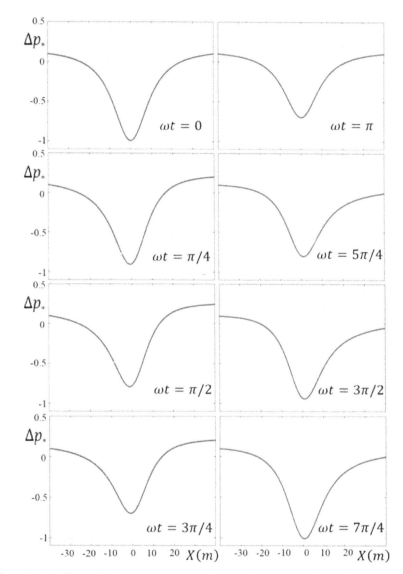

Fig. 10.6. The relative pressure variation in the region of an inhomogeneous train at different times.

of which varies with time. At the initial time, $\Delta p_* = -1$ (in our example, $\rho g A = 0.05 \ p_a$). After half a period, the relative pressure increases to -0.7, but then returns to its original value. At the ends of the inhomogeneity interval, the pressure drop fluctuates in a similar way.

The wave vorticity is equal to

$$\Omega = 2\omega(kA)^2 \left[1 - \beta \frac{2a^2 - 2(b-\alpha)^2}{(a^2 + b^2 + \alpha^2)^2} - (kA)^2 \exp(2kb)\right]^{-1} \exp(2kb).$$

In contrast to the stationary Gerstner wave, the wave vorticity depends on both Lagrangian variables. The properties of the function $|G'|^2$ (the first two terms in square brackets) were studied earlier for $b = 0$. Taking this into account, it can be concluded that the vorticity has a minimum at the point $a = 0$ and maxima at the points $\pm a_*$. In the extreme case (10.15), where the expression in square brackets becomes zero, there are cusps at the points corresponding to the coordinate values $\pm a_*, 0$. They are not of significant interest since they always exist.

References

[1] Philips, O.M. (1977). *The Dynamics of the Upper Ocean* (Cambridge University Press, Cambridge).

[2] Crapper, G.D. (1957). An exact solution for progressive capillary waves of arbitrary amplitude. *J. Fluid Mech.*, 2(4), pp. 532–540.

[3] Leblanc, S. (2007). Amplification of nonlinear surface waves by wind. *Phys. Fluids*, 19, 101705.

[4] Kharif, C., Giovanangeli, J.-P., Touboul, J., Grare, L. and Pelinovsky, E. (2008). Influence of wind on extreme wave events: Experimental and numerical approaches. *J. Fluid Mech.*, 594, pp. 209–247.

[5] Onorato, M. and Proment, D. (2012). Influence of wind on extreme wave events: Experimental and numerical approaches. *Phys. Lett. A.*, 376(45), pp. 3057–3059.

[6] Chabchoub, A., Hoffman, N., Branger, H., Kharif, C. and Akhmediev, N. (2013). Influence of wind on extreme wave events: Experimental and numerical approaches. *Phys. Fluids*, 25, 101704.

[7] Brunetti, M., Marchiando, N., Berti, N. and Kasparian, J. (2014). Nonlinear fast growth of water waves under wind forcing. *Phys. Lett. A.*, 378, pp. 1025–1030.

[8] Eeltink, D., Lemoine, A., Branger, H., Kimmoum, O., Kharif, C., Carter, J., Chabchoub, A., Brunetti, M. and Kasparian, J. (2017). Spectral up- and downshifting of Akhmediev breathers under wind forcing. *Phys. Fluids*, 29, 107103.

[9] Miles, J.W. (1957). On the generation of surface waves by shear flows. *J. Fluid Mech.*, 3, pp. 185–204.

[10] Yan, S. and Ma, Q. (2011). Improved model for air pressure due to wind on 2D freak waves in finite depth. *Eur. J. Mech.* (B Fluids), 30, 1–11.

[11] Kharif, C., Giovanangeli, J.P., Touboul, J., Grare, L. and Pelinovsky, E.N. (2008). Influence of wind on extreme wave events: Experimental and analytical approaches. *J. Fluid Mech.*, 594, pp. 209–247.

[12] Dyachenko, A.I. and Zakharov, V.E. (2008). On the formation of freak waves on the surface of deep water. *JETP Lett.*, 88(5), pp. 307–311.

[13] Talipova, T.G. and Pelinovsky, E.N. (2009). Simulation of a "Lavrenov wave" on the surface of a shallow sea. *Fundamentalnaya i Prikladnaya Gidrofizika*, 2(4), pp. 30–39 [in Russian].

Chapter 11

Vortex Model of Rogue Waves

"The sea also has its own migraines."

— Hugo, V. *The Man Who Laughs*

Rogue waves, also referred to as freak waves, are waves of large amplitude that arise on the sea surface all of a sudden and disappear just as quickly. Their characteristic feature is the amplitude criterion according to which the height of rogue waves is twice the average height, or more, of the surrounding waves [1–4]. Being first considered for ocean waves, the concept of rogue waves has shifted to other fields of physics, such as nonlinear optics, physics of plasma, superfluid helium, and Bose condensate systems. Currently, it is of great interest to elucidate the possible mechanisms of the formation of rogue waves and scenarios for their occurrence in various physical conditions, which ultimately determine the parameters and properties of extreme waves.

The formation of rogue waves is a nonlinear effect [5] studied in a weakly nonlinear approximation within the framework of the nonlinear Schrödinger equation [6–14] and the Dysthe equation [15]. It was found that waves of anomalous amplitude can arise as a result of modulation instability (see reviews [1, 4, 16]). Dyachenko and Zakharov suggested that the focusing of ocean waves only creates preconditions for the formation of rogue waves, which is a strongly nonlinear effect. Having solved a complete system of fluid dynamics equations, they demonstrated that a rogue wave can be formed from a weakly nonlinear Stokes wave [17]. All these theoretical studies

206 *Analytical Fluid Dynamics in Lagrangian Variables*

were carried out with potential wave motion and constant pressure on the free surface of the fluid. These assumptions are justified in the absence of wind. However, rogue waves often occur when the wind impact cannot be neglected. Firstly, the wind changes the pressure on the fluid surface and, secondly, the wave motion becomes vortical. The first factor was taken into account in some works [18–24], where the dynamics of weakly nonlinear, narrow-bandwidth trains of potential surface waves in a field of variable external pressure, defined by the linear theory of wind wave excitation, was studied. We propose a vortex model of rogue waves.

11.1. Birth from the Gerstner Wave

We now consider the Ptolemaic solution of the form [25]

$$W = \chi + \frac{\beta_1}{(\chi - \alpha i)^n} + \left[iA \exp(ik\bar{\chi} + i\varphi_0) + \frac{\beta_2}{(\bar{\chi} + \alpha i)^n} \right] e^{-i\omega t}. \quad (11.1)$$

Here, A, k, ω, and α are positive parameters; $n \geq 2$. When $\beta_1 = \beta_2 = 0$, expression (11.1) describes the Gerstner wave. For Ptolemaic flows, the superposition principle holds true. If the function F is the sum of functions, the resulting profile qualitatively corresponds to the superposition of profiles defined by these functions. The terms in G and F have one pole of order n, which corresponds to $b = \alpha > 0$, so it is outside the flow region. The term with the pole in the function F describes a periodically appearing peak. The term with the pole in the function G compensates for the peak of the wave profile at the initial moment of time. So, expression (11.1) corresponds to the peak standing out in the field of the Gerstner wave.

In solution (11.1), A is the amplitude, ω is the frequency, and k is the wave number of the Gerstner wave; $kA \leq 1$. The equality corresponds to the wave with sharp crests on the profile. The parameter φ_0 characterizes the phase shift between the crests of the Gerstner wave and the vortex breather. If $\varphi_0 = \pi$, their crests coincide and the amplitudes are summarized. If $\varphi_0 = 0$, the breather crest coincides with the trough of the Gerstner wave and their amplitudes are subtracted. This behavior of the solution can be interpreted as wave interference. The parameter α characterizes the peak width.

The values β_1 and β_2 are not necessarily real numbers. To compensate for the peak at $t = 0$ and reinforce it at another moment, let

$$\beta_1 = (-1)^m \beta i; \quad \beta_2 = \bar{\beta}_1, \quad \beta > 0, \quad m = \frac{n}{2}.$$

The value of n should be an even integer number. The dimension of β is L^{n+1}, and its value characterizes the peak height.

We consider a particular case where

$$n = 2, \quad \beta_1 = -\beta i, \quad \beta_2 = \beta i. \tag{11.2}$$

According to (10.2), there is a constraint on the value of β. A sufficient condition is formulated as

$$\beta \leq \frac{(1 - kA)\alpha^3}{4}.$$

The exact bounding for the parameter β can be calculated numerically.

Figure 11.1 represents the dynamics of a freak wave (11.1) for the case $A = 0.5\,m, k = 0.074\,m^{-1}, \alpha = 12\,m$, and $\beta = 328\,m^3$; the wavelength is $\lambda = 84.9\,m$. The phase shift is $\varphi_0 = \pi$. The peak height is $h = 2\beta/\alpha^2 + A \approx 5.1\,m$. At the time $t = 0$, there is no peak and the wave profile corresponds to the Gerstner wave exactly. Next time, the peak rises up to the maximum value at the time $t = \pi/\omega$ and then decreases and disappears after a period. The motion is periodic.

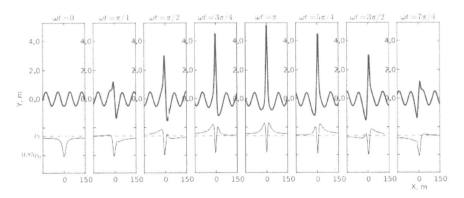

Fig. 11.1. Formation of a rogue wave.

208   *Analytical Fluid Dynamics in Lagrangian Variables*

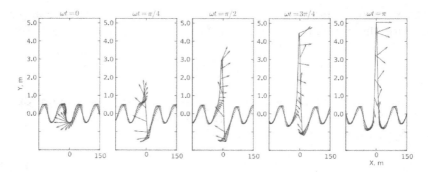

Fig. 11.2. The field of velocities for a rogue wave on the free surface.

The peak height is a factor of eight greater than the amplitude of the Gerstner wave. Thus, such a peak can be considered a rogue wave. The pressure has a non-stationary character. In Fig. 11.1, the lower curves represent the deviation of the pressure on a free surface from atmospheric pressure p_0.

Figure 11.2 shows the velocities of fluid particles at different times. The parameter values are the same as in Fig. 11.1. During the formation of rogue waves, the velocities of the front and back slopes of the peak have opposite directions (see the times $t = \pi/4\omega$ and $t = \pi/2\omega$). So, the wave profile collapses and rises. When the velocity at the highest point becomes horizontal, the rogue wave starts to decrease.

The flow vorticity given by Eqs. (11.1) and (11.2) is equal to

$$\Omega = \frac{2\omega |F'|^2}{|G'|^2 - |F'|^2};$$

$$|G'|^2 = 1 + \frac{4\beta^2 + 4\beta(b-\alpha)[3a^2 - (b-\alpha)^2]}{[a^2 + (b-\alpha)^2]^3}; \qquad (11.3)$$

$$|F'|^2 = (kA)^2 e^{2kb} + \frac{4\beta^2 - 4\beta kAe^{kb}\mathrm{Im}\{e^{ika}[a - i(b-\alpha)]^3\}}{[a^2 + (b-\alpha)^2]^3}.$$

Far away from the peak location, the value of Ω is near the vorticity of the Gerstner wave. For the Gerstner wave of weak steepness ($kA \ll 1$), the flow outside the rogue wave is almost potential. The vorticity localizes in the small neighborhood of the peak location. Thus, the rogue wave is strongly vortical.

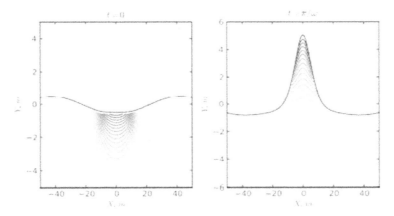

Fig. 11.3. Isolines of vorticity.

Figure 11.3 represents the isolines of vorticity (Ω = const in (11.3)) in the neighborhood of the peak location at two moments of time. The light gray isolines correspond to the smaller vorticity and the dark gray ones to the greater vorticity. With the lapse of time, the isolines of vorticity become convex near the peak axis. So, the formation of a rogue wave in our model is due to the bending of vorticity isolines.

As is seen in Fig. 11.1, the minimum pressure on the free surface corresponds to the particle $a = b = 0$. Because of the non-stationarity of the flow, we estimate the pressure at two qualitatively different moments of time. At the moment $t = 0$, when there is no peak, the pressure is $p(0)$ and at the moment $t = \pi/\omega$, when the peak height is the maximum, the pressure is $p(\pi)$. The case $p(0) = p(\pi) = -100\,mm$ of mercury (mm Hg) is studied.

Table 11.1 represents examples of rogue wave parameters obtained by numerical calculations. Here, $\lambda = 2\pi/k$ is the length of the Gerstner wave. If the steepness is near 1, then the value of h tends to 0. The peak cannot form on a steep wave. The ratio between the height of the peak and Gerstner wave amplitude is $(1/A - k)\alpha/2$. It can be very large. With a small steepness, the maximum waves are located in the wavelength range $\lambda \sim 20,\ldots,60\,m$.

Our estimates of the wave parameters are given for a pressure deviation of the order of $|p(0)| = 100\,\text{mm Hg}$. The magnitude of the pressure gradient can vary depending on the breather width, which is of the order of α. Table 11.2 represents a relation between the

Table 11.1. Parameters of maximal freak wave.

A, m	k A	λ, m	h, m
0.25	0.07	21.2	4.8
0.5	0.13	23.5	5.4
2.0	0.33	38.0	8.8
4.0	0.45	56.1	12.3
1.25	0.69	11.4	1.8
1.25	0.75	10.5	1.4
1.25	0.88	9.0	0.7

Table 11.2. Wave parameters for given amplitude A.

A, m	kA	α, m	h, m	$|p(0)|/\alpha$,mm Hg/m
4.0	0.44	15.0	7.6	6.6
4.0	0.40	20.0	6.0	5.0
4.0	0.33	30.0	4.7	3.3
4.0	0.27	40.0	4.2	2.5

average pressure gradient $|p(0)|/\alpha$ and the value of α for a given Gerstner wave amplitude $A = 4\,m$. These conditions do not correspond to the maximum rogue wave. When the pressure drop occurs on a larger scale (fat high values of α), the height of the vortex breather decreases. The considered values of pressure, pressure gradient, and the breather width correspond to tornadoes.

Modulation instability is considered one of the most probable mechanisms for the formation of extreme waves. However, our mechanism for the formation of anomalous waves is significantly different. It is based on the non-uniform distribution of pressure over the fluid surface. Thus, the pressure gradient plays the role of an external force (see also [26]).

11.2. Special Pressure Mode

Our vortex model is based on an exact analytical solution of the equations of 2D fluid dynamics (Ptolemaic flows). A unique feature of flows of this class is the dependence of the coordinates of motion of fluid particles on two complex functions, which can be largely arbitrary. As a consequence, the model can be used for the analysis of various representations of surface pressure, as well as fluid vorticity, Both of these factors characterize the effect of wind on the free surface of water. An exact solution for which the surface pressure varied in antiphase with the free surface elevation was studied in previous section. However, Phillips [27] emphasized that the phase difference between fluctuations in surface pressure and wave profile in the natural environment can be very diverse, and statistical sampling of wave observations does not clearly favor any value or range of values.

In Section 11.1, we observed a pressure pit relative to the constant (atmospheric) level p_0 at $t = 0$. The order of magnitude of the pit was equal to α. Then, it started to narrow, while its depth remained almost constant. Two peaks above the p_0 level were formed on the left and right of the absolute pressure minimum. Half a period later, they reached their maximum (smaller than the pit depth). The rogue wave amplitude was maximum at this moment of time. It was decreasing gradually during the next half-period, and the pressure distribution recovered the form it had at the initial moment of time. During the oscillation period, the pressure changed significantly.

In the next section, we consider a special pressure distribution that differs from the above-mentioned example. This distribution has a negative pressure pit so that the peak of the wave profile is first in phase with the pressure and then antiphase. In this fashion, we simulate a qualitatively self-consistent behavior of the wave profile and the pressure on it typical for oceanic and laboratory conditions [27]. The form of the exact solution is chosen such that the pressure should be time-independent in Lagrangian variables [28, 29].

11.2.1. *Stationarity of pressure in the Ptolemaic flow*

We now address surface waves for which pressure (10.4) does not depend on time. This means that the functions F and G meet the following condition:

$$g\bar{F} + i\omega^2 \int G'\bar{F}d\chi = 0.$$

212   *Analytical Fluid Dynamics in Lagrangian Variables*

Differentiation of this equality with respect to χ yields

$$G' = \frac{ig}{\omega^2}\frac{\bar{F}'}{\bar{F}}.\tag{11.4}$$

By integrating (11.4), we find

$$G = \frac{ig}{\omega^2}\ln\frac{\bar{F}}{\alpha},\tag{11.5}$$

where α is a real constant of dimensional length (the value under ln sign is dimensionless). Here, F is an arbitrary analytical function that has no singularities in the flow region $\operatorname{Im}\chi \leq 0$. The function G is found by the known form of F from expression (11.5). The constant α specifies the horizontal scale of F variation. An additional requirement for choosing functions G and F is the condition of a non-negative Jacobian (10.2). Making use of equality (11.4), this condition can be transformed to

$$\left|\frac{g}{\omega^2}\frac{F'}{F}\right| \geq |F'|.\tag{11.6}$$

The function F is represented in the form

$$F(\bar{\chi}) = iA(1 + \overline{P(\chi/\alpha)})\exp(ik\bar{\chi}),\tag{11.7}$$

where A and k are the amplitude and wave number of the wave motion, respectively, and P is an analytical function. For $P = 0$, expressions (11.5) and (11.7) correspond to the Gerstner wave. The pressure on the Gerstner wave profile is constant. Hereinafter, the function P will be chosen as a localized perturbation with horizontal scale α. At fairly far distances from the perturbation ($|\mathrm{Re} \gg \alpha|$), the solution of Eqs. (10.1), (11.5), and (11.7) will transform to the Gerstner wave solution. Hence, the value of the wave number should be chosen as in the Gerstner wave ($k = \omega^2/g$).

Taking the above-mentioned information into account, inequality (11.6) is equivalent to the fulfillment of two conditions in the region

$$F'(\bar{\chi}) \neq 0,\quad |kF(\bar{\chi})| = |kA(1 + \overline{P(\chi/\alpha)})| \leq 1,\quad \operatorname{Im}\chi \leq 0.\tag{11.8}$$

The first condition demands that F (and, hence, P) be a single-valued function and the second limits the magnitude of the wave

Vortex Model of Rogue Waves

213

perturbation amplitude. For the Gerstner wave $(P = 0)$, it transforms to the inequality $kA \leq 1$. The case of equality corresponds to the wave-limiting amplitude $A = k^{-1}$. The wave crest in this case coincides with the profile cusp (at this point, the tangent to it is vertically directed).

Now, we can write partial form of the studied solution to (10.1) as

$$W = \chi + \frac{i}{k} \ln[1 + P(\chi/\alpha)] + iA(1 + \overline{P(\chi/\alpha)}) \exp i(k\bar{\chi} - \omega t). \quad (11.9)$$

This expression specifies a family of exact solutions dependent on one function P only. It must be analytical and bounded on the real axis, must meet condition (11.8), and must not take on the value -1 in the flow region (to restrict the logarithmic term in (11.9)).

The first two terms in (11.9) determine an average level of fluid. The function $A(1 + \overline{P(\chi/\alpha)})$ is the amplitude of the Gerstner wave train envelope. The case $\alpha \gg \lambda = 2\pi/k$ corresponds to the wave train with slowly varying amplitude. Such wave motions are intensely studied within the framework of the nonlinear Shrödinger equation for the amplitude of the wave train envelope under the condition of a potential flow and constant pressure on the free surface. However, we are interested in the dynamics of a strongly nonlinear wave train when $\alpha \ll \lambda$. It corresponds to a breather evolving against the background of uniform Gerstner waves.

The pressure on the free surface is defined by

$$\frac{p - p_0}{\rho g} = \frac{k}{2}|F|^2 - \frac{1}{k} \ln \left| \frac{F}{\alpha} \right|, \quad \text{Im} \, \chi = 0, \quad (11.10)$$

where F is given by equality (11.7). The value of the constant p_0 is taken to ensure the pressure on the free surface of a purely Gerstner wave (at $\text{Re} \, \chi \to \pm\infty$ and, hence, $P \to 0$) equal to the atmospheric pressure p_a. Then, expression (11.10) will be rewritten in the form

$$\frac{p - p_a}{\rho g} = \frac{k}{2}(|F|^2 - A^2) - \frac{1}{k} \ln |F/A|; \quad \text{Im} \, \chi = 0. \quad (11.11)$$

Recall that the pressure is time-independent only in Lagrangian coordinates, whereas it is a function of time in the Euler description.

The expression for flow vorticity (11.9) will be written as

$$\Omega = \frac{2\omega k^2 |A(1+\bar{P})|^2 \exp(2kb)}{|G'|^2 - k^2|A(1+\bar{P})|^2 \exp(2kb)},$$

$$|G'|^2 = 1 + \frac{1}{(k\alpha)^2}\frac{|P'|^2}{(1+P)(1+\bar{P})} - \frac{2}{k\alpha}\text{Im}\frac{P'}{1+P}.$$

(11.12)

In the absence of localized perturbation, when $P = 0$, the value of Ω is equal to the Gerstner wave vorticity.

11.2.2. *Quasi-stationary pressure pit*

Let us take the function P in the form

$$P(\chi/\alpha) = \frac{i\beta}{i\alpha - \chi},$$

(11.13)

where β is a positive constant having the dimension of length. This function has a pole at the point $\text{Im}\,\chi = \alpha$, but due to the positivity of α, this point is located outside the flow region. Similarly, the point $\chi = i(\beta + \alpha)$, where P is equal to -1, corresponds to the region $\text{Im}\,\chi > 0$ and is located above the fluid level.

Let us assume $A > 0$. Then, the condition $F' = 0$ reduces to the equation

$$k(\bar{\chi} + \alpha)^2 + ik\beta(\bar{\chi} + \alpha) - \beta = 0,$$

(11.14)

the roots of which lie outside the flow region, i.e., in the half-plane $\text{Im}\,\bar{\chi} < 0$. Assuming $\bar{\chi} = iw$, we rewrite Eq. (11.14) in the form

$$kw^2 + k(2\alpha + \beta)w + k\alpha(\alpha + \beta) + \beta = 0.$$

As all the coefficients of this quadratic equation are positive, according to the Routh–Hurwitz criterion, its roots must meet the condition $\text{Re}\,w = \text{Im}\,\bar{\chi} < 0$. Consequently, for positive values of k, A, α, β, the function F' vanishes to zero only at the points corresponding to the inequality $\text{Im}\,\chi > 0$.

The function $1 + P$ is analytical; hence, it achieves its maximum value at the boundary of the flow region, where $\text{Im}\,\chi = 0$. The inequality $|1 + P| \leq 1 + (\beta/\alpha)$ holds for the absolute value of this function. With allowance for this inequality, (11.8) will be written in the following form:

$$kA[1 + (\beta/\alpha)] \leq 1. \tag{11.15}$$

The quantity β/α has an upper limit equal to $(1/kA) - 1$. If the Gerstner wave steepness kA is close to zero, then β/α may have a very large value.

Figure 11.4 shows the breather evolution and the pressure distribution on the free surface at different times for the following values of the parameters:

$$A = 2.5m, \quad \alpha = 1m, \quad \lambda = 30m, \quad k = 0.21m^{-1},$$
$$\omega = \sqrt{gk} = 1s^{-1}, \quad \beta = 0.9m.$$

The minimum pressure is $100\,\mathrm{mm\,Hg}$ lower than the atmospheric pressure. The initial moment of time is $t_1 = \pi/\omega$.

The formation of a rogue wave starts in the Gerstner wave trough. A specific feature of this process is shown on a magnified scale in the inset on the left of the Figure 11.4. First, two local maxima start to grow in the trough in the regions corresponding to the edges of the pressure pit. This leads to the conclusion that the force of the pressure gradient plays a decisive role in the initial stage of wave evolution. Later, at $t_2 = 5\pi/4\omega$, only the left maximum (for which $X < 0$) remains, with the geometric center shifted to the right, closer to the origin of coordinates. At $t_3 = 3\pi/2\omega$, its height already exceeds the Gerstner wave amplitude; the peak over the "amplitude level" has a non-symmetric shape. At the next time $t_4 = 7\pi/4\omega$, the amplitude of the peak increases, and it becomes more symmetric. Finally, at $t_5 = 2\pi/\omega$, the wave height reaches its maximum, and its profile becomes symmetric to the vertical $X = 0$. During the next half-period, the height of the peak decreases monotonically, and its center shifts to the region of increasingly positive X. The wave profile evolves similarly to the previous half-period: The patterns at $t_6 = 9\pi/4\omega$ and t_4, $t_7 = 5\pi/2\omega$ and t_3, and $t_8 = 11\pi/4\omega$ and t_2 are mirror symmetric and transform one into another upon replacement of $X \to -X$.

At the bottom of Fig. 11.4, pressure is plotted as a function of the horizontal coordinate X at the corresponding times. It is seen that the pressure deforms over time. It narrows during a half-period and then widens to its previous state. An important property of the

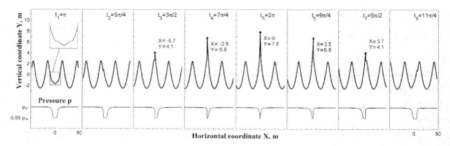

Fig. 11.4. Dynamics of the profile (upper curves) and pressure (lower curves) on the free surface over one wave period. Numerical values for horizontal X and vertical Y coordinates are given for the wave crest.

pressure drop is its negative value. We call the model with such pressure behavior quasi-stationary to distinguish it from the case where the deviation of pressure from the atmospheric level is an alternating non-stationary function of time (see Section 11.1). Unlike the non-stationary model, the quasi-stationary model explicitly demonstrates the mechanism of the rogue wave formation: The pressure pit is compressed along the horizontal and the fluid is forced to the maximum possible height.

This mechanism for the formation of extreme waves is not connected to modulation instability. The solution that arises as a result of modulation instability can be called free, while our solution belongs to the class of forced ones.

We consider a periodic solution when a rogue wave forms in a short time and disappears just as quickly. In our model, this is equivalent to the fact that a "pit" of external surface pressure, generating an anomalous wave, is formed and exists at times of the order of the wave period. Large pressure drops are required to form sufficiently high waves. For numerical simulations, we chose a pressure drop of 100 Hg mm. Such pressure deviations are rare events that occur only during strong wind. Besides, our model assumes that these events are short-lived.

References

[1] Dysthe, K., Krostad, H.E. and Müller, P. (2008). Oceanic rogue waves. *Ann. Rev. Fluid Mech.* 40, pp. 287–310.

[2] Osborne, A.R. (2009). *Nonlinear Ocean Waves* (Academic Press).

[3] Kharif, C., Pelinovsky, E., Slunyaev, A. (2009). *Rogue Waves in the Ocean* (Berlin: Springer-Verlag).

[4] A. Slunyaev, I. Didenkulova and E. Pelinovsky. (2011) Rogue waves. *Contemp. Phys.*, 52(6), 571–590.

[5] Ruban, V., Kodama, Y., Ruderman, M., Dudley, J., Grimshaw, R., McClintock, P.V.E. *et al.* (2010). Rogue waves — towards a unifying concept? Discussions and debates. *Eur. Phys. J. Spec.Top.*, 185(1), pp. 5–15.

[6] Peregrine, D.H. (1983). Water waves, nonlinear Schrödinger equations and their solutions. *J. Aust. Math. Soc. Ser. B.*, 25, pp. 16–43.

[7] Ankiewicz, A., Kedziora, D. and Akhmediev, N. (2011). Rogue wave triplet. *Phys. Lett. A.*, 375(28), pp. 2782–2785.

[8] Ankiewicz, A., Devine, N. and Akhmediev, N. (2009). Are rogue waves robust again perturbations? *Phys. Lett. A.*, 373(43), pp. 3997–4000.

[9] Onorato, M., Osborne, A.R. and Serio, M. (2006). Modulational instability in crossing sea states: A possible mechanism for the formation of freak waves. *Phys. Rev. Lett.*, 96, 014503, 5p.

[10] Dyachenko, A.I. and Zakharov, V.E. (2008). On the formation of freak waves on the surface of deep water. *JETP Lett.*, 88, pp. 307–311.

[11] Toffoli, A., Bitner-Gregersen, E.M., Osborne, A.R., Serio, M., Monbaliu, J. and Onorato, M. (2011). Extreme waves in random crossing seas: Laboratory experiments and numerical simulations. *Geophys. Res. Lett.*, 38, L06605, 5p.

[12] Chabchoub, A. Hoffman, N.P. and Akhmediev, N. (2011). Rogue wave observation in a water wave tank. *Phys. Rev. Lett.*, 106, 204502, 4p.

[13] Kedziora, D.J., Ankiewicz, A. and Akhmediev, N. (2013). Classifying the hierarchy of nonlinear-Schrödinger-equation rogue-wave solutions. *Phys Rev E.*, 88(1), 013207, 12p.

[14] Ablowitz, M.J. and Horikis, T.P. (2015). Interacting nonlinear wave envelopes and rogue wave formation in deep water. *Phys. of Fluids.*, 27, 012107, 10p.

[15] Ablovitz, M.J., Hammack J., Henderson, D. and Scholder, S.M. (2000). Modulated periodic Stokes waves in deep water. *Phys. Rev. Lett.*, 84, pp. 887–890.

[16] Kharif, C. and Pelinovsky, E. (2003). Physical mechanisms of the rogue wave phenomenon. *Eur. J. Mech. B, Fluids*, 22, pp. 603–634.

[17] Dyachenko, A.I. and Zakharov, V.E. (2005). Modulation instability → freak wave. *JETP Lett.*, 81(6), pp. 255–259.

[18]   Leblanc, S. (2007). Amplification of nonlinear surface waves by wind. *Phys. Fluids,* 19(10), 101705

[19]   Kharif, C., Giovanangeli, J.-P., Touboul, J., Grare, L. and Pelinovsky, E. (2008). Influence of wind on extreme wave events: Experimental and numerical approaches. *J. Fluid Mech.,* 594, pp. 209–247.

[20]   Adcock, T.A.A. and Taylor, P.H. (2011). Energy input amplifies nonlinear dynamics of deep water wave groups. *Int. J. Offsh. Polar Eng.,* 21(1), pp. 8–12.

[21]   Onorato, M. and Proment, D. (2012). Approximate rogue wave solutions of the forced and damped nonlinear Schrödinger equation for water waves. *Phys. Lett. A.,* 376(45), pp. 3057–3059.

[22]   Chabchoub, A., Hoffman, N., Branger, H., Kharif, C. and Akhmediev, N. (2013). Experiments on wind-perturbed rogue wave hydrodynamics using the Peregrine breather model. *Phys. Fluids,* 25(10), 101704, 4p.

[23]   Brunetti, M., Marchiando, N., Berti, N. and Kasparian, J. (2014). Nonlinear fast growth of water waves under wind forcing. *Phys. Lett. A.,* 378(14), pp. 1025–1030.

[24]   Kharif, C., Kraenkel, R.A., Manna, M.A. and Thomas, R. (2010). The modulational instability in deep water under the action of wind and dissipation. *J. Fluid Mech.,* 664, pp. 138–149.

[25]   Abrashkin, A.A. and Soloviev, A.G. (2013). Vortical freak waves in water under external pressure action. *Phys. Rev. Lett.,* 110, 014501, 4p.

[26]   Abrashkin, A.A. and Soloviev, A.G. (2013). Gravity waves under nonuniform pressure over a free surface. Exact solutions. *Fluid Dynamics,* 48(5), pp. 679–686

[27]   Philips, O.M. (1977). *The Dynamics of the Upper Ocean* (Cambridge University Press, Cambridge).

[28]   Abrashkin, A.A. and Oshmarina, O.E. (2016). Rogue wave formation under the action of quasi-stationary pressure. *Comm. Nonl. Sci. Num. Simul.,* 34, pp. 66–76

[29]   Abrashkin, A.A. and Oshmarina, O.E. (2014). Pressure induced breather overturning on deep water: Exact solution. *Phys. Lett. A.,* 378, pp. 2866–2871.

Chapter 12

Breaking of the Surface Gravity Wave

"The waves are running, one
breaking another's crest..."

— O. Mandelstam

The breaking of surface gravity waves on water is one of the brightest examples of their nonlinear behavior. Two types of breaking waves, namely, spilling and plunging breakers, may be distinguished [1, 2]. The spilling breaker slowly gets steeper. When its steepness reaches some critical value, the water at the crest begins to slide ("spill") down the wavefront, reducing the wave height. Thus, the wave's steepness does not exceed this critical value. The plunging breaker becomes steep very quickly. Its final steepness is greater than that of the spilling breaker. The top of the plunging wave gets far ahead of the bottom of the wave, thus forming an arc-shaped profile. Mathematically, this corresponds to the appearance of a vertical tangent of a free surface profile, which is the basic difference between the two wave types. In this chapter, we are interested in plunging breakers.

This phenomenon is well studied in shallow-water approximation [3, 4]. The situation with deep water is less clear. Full information about the wave profile, velocity, and pressure fields for breaking waves was obtained by complex numerical calculations [5–7]. But there are also some analytical results. Using the semi-Lagrangian approach proposed by John [8], Longuet-Higgins [9,10] and New [11] developed

220   *Analytical Fluid Dynamics in Lagrangian Variables*

simple analytical models to determine the underside or loop of a plunging breaking wave. Longuet-Higgins suggested that the solution for a rotating hyperbola falling under gravity could represent the motion of a jet of fluid ejected from the top of the breaking wave. New found that the free surface under the jet is well described by an ellipse and obtained unsteady solutions for a zero-gravity flow around the ellipse. Greenhow [12] showed that both the fluid jet from the top of the breaking wave and the ellipse model describing the loop are complementary solutions to the same free surface equation for large times. This, in turn, allowed us to propose solutions combining both the jet and the loop and to obtain a much more complete model of the entire breaking region near the wave crest.

All these theoretical studies were carried out for potential flows. The pressure at the free boundary was assumed to be constant. But most waves in the ocean occur in storm conditions, where the wind action must be taken into account. As in the previous two chapters, we will relate the wind action to the non-uniform and unsteady pressure acting on the free surface. In doing so, we will abandon the wave potential approximation.

12.1.   Deep-Water Breather Dynamics

Let us consider the following Ptolemaic solution:

$$W = \chi - \frac{i\beta}{(\chi - i)^2} + \frac{i\beta}{(\bar{\chi} + i)^2} e^{-i\omega t}, \tag{12.1}$$

where β is a real value, $\mathrm{Im}\,\chi \leq 0$, and ω is the oscillation frequency. The functions G and F (see Eq. (10.1)) have a pole at the point $\mathrm{Im}\,\chi = 1$, which is outside the flow region. The solution is dimensionless. The values of W and χ are normalized on the length scale α, and the parameter β is normalized by α^3. The properties of the breather are determined only by the constant β. Let us study the range of its admissible values.

The function G' should not vanish in the flow region. Thus, the equation $(\chi - i)^3 + 2i\beta = 0$ should have roots with $\mathrm{Im}\,\chi > 0$ only. Using the substitution $\chi = -iz$, we obtain a cubic equation in the following form:

$$(z + 1)^3 + 2\beta = 0, \quad \mathrm{Re}\,z < 0. \tag{12.2}$$

According to the Routh–Hurwitz criterion, Eq. (12.2) has roots with negative real parts if and only if $-0.5 < \beta < 4$.

The Jacobian (10.2) must be positive, i.e.,

$$\left| 1 + \frac{2\beta i}{(\chi - i)^3} \right|^2 > \left| \frac{2\beta i}{(\bar{\chi} + i)^3} \right|^2 = \left| \frac{2\beta i}{(\chi - i)^3} \right|^2 .$$

This condition is equivalent to the inequality $|(\chi - i)^3 + 2\beta i|^2 > 4\beta^2$. Its left-hand side contains an analytical complex function. The modulus of this function reaches its minimum value at the boundary of the region corresponding to the free surface, where $b = 0$. Considering the extrema of the function $|(a - i)^3 + 2\beta i|^2$, we find the admissible range of β: $-0.25 < \beta < 1$.

Solution (12.1) has a qualitatively different evolution depending on the value of β. We start the analysis with $t_0 = \pi/\omega$. At this point in time, the shape of the free boundary is symmetric relative to the vertical axis passing through the maximum elevation point (see Fig. 12.1 (a)). Its height is equal to 2β (or $2\beta a$ in dimensional form). When β is negative, the free surface profile has a trough. At $\beta > 0$, the profile has a crest.

Figure 12.1 shows the evolution of the breather during one period of oscillation. The free motion perturbation changes shape but does not move as a whole. Therefore, we can consider it to be a breather. Eventually, the breather profile tilts. Two qualitatively different regimes are possible: (i) for $\beta = -0.25$ or $\beta = 0.5$, the profile has no contraflexure points and (ii) for $\beta = 0.85$, the profile has contraflexure points, and we see that the breather breaks.

We focused on the profile with $\beta = 0.85$ depicted by a bold curve. The breather steepness increases with time, and at $t_1 = 4.25/\omega$ (Fig. 12.1(b)), a contraflexure point appears for the first time. This point is characterized by a vertical tangent to the profile, depicted by a dotted line. The profile then has two contraflexure points until the moment $t_3 = 5.9/\omega$, when they merge. Then, the contraflexure point disappears. At $t = 2\pi/\omega$, the surface becomes flat.

During the next half-cycle, all stages of the profile evolution are repeated symmetrically (see Figs. 12.1(e)–(g)), but contraflexure points occur on the left slope of the breather. A contraflexure point arises at $t_4 = 6.7/\omega$, when the breather steepness is small enough. This is unlikely to be observed under natural conditions. But the

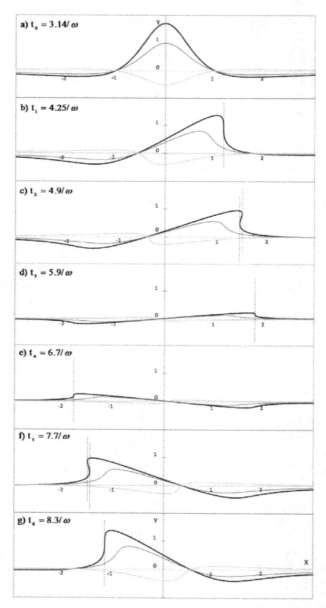

Fig. 12.1. Dynamics of the breather profile at different values of β. The bold black line corresponds to $\beta = 0.85$, the black line to $\beta = 0.5$, and the gray line to $\beta = -0.25$.

Breaking of the Surface Gravity Wave

dynamics of the breather during the first half-cycle is very similar to the breaking of an ocean wave. Close to the moment when the vertical tangent appears, the viscosity of the fluid should be taken into account. Its action leads to the wave breaking at some moment $t_* > t_1$. Thus, Eq. (12.1) describes the overturning of real waves on water within the time interval (t_0, t_*). For example, we can consider $t_* = t_2$.

Solution (12.1) corresponds to non-uniform pressure on the free surface. Figure 12.2 shows comparative oscillograms of the profile and pressure simultaneously for $\beta = 0.85$. The upper curve corresponds to the profile. The dashed circles represent the trajectories of the fluid particles. Bold dots indicate the position of some fluid particles on the free surface.

Following Eq. (12.1), the fluid particles rotate in circles of different radii with frequency ω. The breather starts to break at time t_1, when two neighboring points of the profile for the first time tend to be located on the same vertical line.

The lower curve in each figure represents the qualitative shape of the pressure curve. Substituting solution (12.1) into (10.4), we find

$$p = p_0 + \rho g \alpha \left\{ \beta P_1(a,t) + \frac{\omega^2 \alpha}{g} \left[\beta P_2(a,t) + \beta^2 P_3(a,t) \right] \right\};$$

$$P_1(a,t) = \frac{(a^2 - 1)(1 - \cos \omega t) + 2a \sin \omega t}{(a^2 + 1)^2};$$

$$P_2(a,t) = \frac{a \sin \omega t - 1}{a^2 + 1};$$

$$P_3(a,t) = \frac{1}{2(a^2 + 1)^2} \left[1 - \frac{(a^2 - 1)^2 - 4a^2}{(a^2 + 1)^2} \cos \omega t \right.$$

$$\left. - \frac{4a^2(a^2 - 1)}{(a^2 + 1)^2} \sin \omega t \right].$$

In contrast to the profile, the pressure p depends on the non-dimensional parameter $\omega^2 \alpha / g$. The value of α determines the horizontal scale of the breather. In numerical calculations, we take $\alpha = 1m$, $\omega = 1s^{-1}$ such that $\omega^2 \alpha / g \ll 1$. In this case, the pressure oscillates in the phase inverse to the profile and corresponds to the breather profile in accordance with the "inverted barometer" rule [13].

224   *Analytical Fluid Dynamics in Lagrangian Variables*

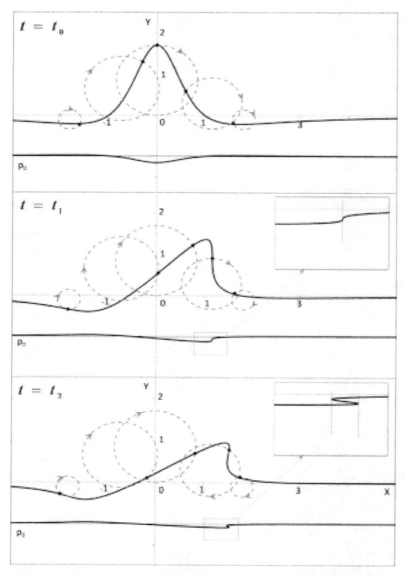

Fig. 12.2. Breather profile and pressure on the free surface (bold curves) and the trajectories of fluid particles (dotted lines). The arrows indicate the direction of their rotation.

Breaking of the Surface Gravity Wave

At the initial moment of time $t = t_0$, the modulus of pressure deviation from atmospheric pressure is

$$|p - p_0| = \rho\alpha\beta(2g + \omega^2\alpha) > 2\rho\alpha\beta g = \rho g H_{br},$$

where H_{br} is the maximum breather height. For example, if $H_{br} = 1m$, then the pressure difference should be more than $70\,mm$ Hg. Such pressure deviations are quite large and are observed only in very strong winds. Thus, solution (12.1) has a limited physical application. It describes the dynamics of not very high breathers.

12.2.   The Breather Breaking Mechanism

The breaking regime is realized in a certain range of β values. Breather breaking is connected to the existence of a vertical tangent to the breather profile. This condition can be written as follows:

$$X^{'}(a) = 0, \quad Y^{'}(a) \neq 0. \tag{12.3}$$

Since the Ptolemaic flow is periodic, wave breaking is also repeated periodically. Condition (12.3) is equivalent to the inequality

$$\left| \mathrm{Re}\, G^{'}(a) \right| \leq \left| F^{'}(a) \right| \tag{12.4}$$

for some values of the Lagrangian coordinate a.

Let β be positive. The value of β is limited to unity. It is easy to find that $\mathrm{Re}\, G^{'}(a) > 0$ for $0 < \beta < 1$. Substituting Eq. (12.1) into Eq. (12.4) yields

$$2\beta \geq \frac{r^6}{r^3 + 3r^2 - 4}, \quad r = \sqrt{a^2 + 1}.$$

The extremum condition for the function on the right-hand side has the following form:

$$r^3 + 4r^2 - 8 = 0.$$

This equation has the roots $r_1 = -2$ and $r_{2,3} = -1 \pm \sqrt{5}$. The only root satisfying the condition $r \geq 1$ is $r = -1 + \sqrt{5}$. This corresponds to the minimum value of $\beta(r)$ when wave breaking is observed. This value is $\beta_* = 0.721$. In this case, the contraflexure point exists for a very short time. Thus, breather breaking takes place with β in the range $0.721 < \beta < 1$. The greater the β, the longer the contraflexure points are observed.

According to Eq. (1.18), the longitudinal gradient of the pressure on the free surface is

$$\left.\frac{\partial p}{\partial X}\right|_{b=0} = \left.\frac{p_a}{X_a}\right|_{b=0} = \left. -\rho\left[X_{tt} + (Y_{tt} + g)\frac{Y_a}{X_a}\right]\right|_{b=0}.$$

The accelerations of the fluid particles X_{tt} and Y_{tt} are bounded. The pressure oscillogram has contraflexure points if condition (12.3) is satisfied, i.e., at the same time as the breather profile. Thus, breather breaking can be interpreted as the action of pressure of a special type.

The breaking phenomenon is explained in Fig. 12.5. The values of all breather parameters are the same as in Fig. 12.2. The arrows correspond to the velocity vectors. On the front slope, the particles move with a higher horizontal velocity near the crest; therefore, the profile becomes twisted. This effect is the basis of the breaking mechanism.

The flow vorticity in Eq. (12.1) is determined by

$$\Omega = \frac{8\beta^2\omega}{[a^2 + (b-1)^2]^3 + 4\beta(b-1)[3a^2 - (b-1)^2]}.$$

The vorticity isolines in the vertex vicinity at three time points are presented in Fig. 12.3. The light gray isolines correspond to smaller vorticity and the dark black ones to greater vorticity. The points of vorticity maximum on the free surface correspond to $a_{1,2} = \pm\sqrt{2\sqrt{\beta} - 1}$, i.e., to the m_1 and m_2 points in Fig. 12.3. If $0 < \beta < 0.25$, then the coordinate of the fluid particle with maximum vorticity is $a = 0$.

There are two types of vortex isolines at the initial moment of time t_0. Isolines of the first type have a concave curved shape. They are located near the breather slopes and correspond to the regions of maximum vorticity. These isolines exist if $\beta > 0.25$. Isolines of the second type have a convex curved shape. The boundary isoline between these two families passes through the top of the breather. All convex isolines are located below it. Eventually, their front slope becomes steeper. The concave isolines around the point m_1 retain their qualitative shape. But the isolines around the point m_2 change notably. They rotate and bend so that, for example, at $t = t_2$, an isoline with two vertical tangents takes place.

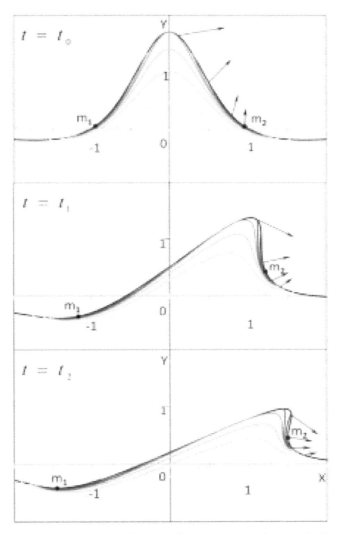

Fig. 12.3. Velocity vectors in the zone of overturning and vortex isolines. Darker lines correspond to higher vorticity. The points m_1 and m_2 denote the maximum vorticity isolines.

Breaking occurs in the maximum vorticity zone. The vorticity value at the point m_2 is equal to

$$\Omega_{m_2} = \frac{\beta \omega}{2(1 - \sqrt{\beta})}.$$

228   *Analytical Fluid Dynamics in Lagrangian Variables*

This expression is an increasing function of β. Thus, we can conclude that breather breaking (12.1) occurs when the vorticity maximum exceeds $\Omega_{m_2}(\beta_*) = 2.4\omega$.

References

[1]   Banner, M.L. and Peregrine, D.H. (1993). Wave breaking in deep water. *Annu. Rev. Fluid Mech.*, 25, pp. 373–397.

[2]   Banner, M.L. and Peirson, W.L. (2007). Wave breaking onset and strength for two-dimensional deep-water wave groups. *J. Fluid Mech.*, 585, pp. 93–115.

[3]   Peregrine, D.H. (1987). Waves on beaches. Numerical study of three-dimensional overturning waves in shallow water. *Annu. Rev. Fluid Mech.*, 15, pp. 149–178.

[4]   Guyenne, P. and Grill, S.T. (2006). *J. Fluid Mech.*, 547, pp. 361–388.

[5]   Longuet-Higgins, M.S. and Cokelet, E.D. (1976). The deformation of steep surface waves on water. I. Numerical method of computation. *Proc. Roy. Soc. London. Ser. A*, 350, pp. 1–26.

[6]   New, A., McIver, P. and Peregrine, H.D. (1980). Computations of overturning waves. *J. Fluid Mech.*, 150, pp. 233–251.

[7]   Dyachenko, A.I. and Zakharov, V.E. (2005). Modulation instability of Stokes wave, *JETP Lett.*, 81(6), pp. 255–259.

[8]   John, F. (1953). Two-dimensional potential flows with a free boundary. *Comm. Pure Appl. Math.*, 6, pp. 497–503.

[9]   Longuet-Higgins, M.S. (1980). On the forming of sharp corners at a free surface. *Proc. Roy. Soc. London A*, 371, pp. 453–478.

[10] Longuet-Higgins, M.S. (1982). Parametric solutions for breaking waves. *J. Fluid Mech.*, 121, pp. 403–424.

[11] New, A.L. (1983). A class of elliptical free-surface flows. *J. Fluid Mech.*, 130, pp. 219–239.

[12] Greenhow, M. (1983). Free-surface flows related to breaking waves. *J. Fluid Mech.*, 134, pp. 259–275.

[13] Wunsch, C. and Stammer, D. (1997). Atmospheric loading and the oceanic "Inverted Barometer" effect. *Reviews of Geophysics*, 35, pp. 79–107.

Chapter 13

Non-Stationary Edge Waves

"Sea breeze drives the waves,
Digs the shoal, argues against the land..."

— V. Bryusov

Edge waves are waves on the water surface that run along the shore. These motions, which have the largest amplitude on the shoreline, usually propagate along the shoreline, but can also form a standing structure with a longshore periodic sequence of high and low amplitudes. Edge waves are also responsible for the formation of beach cusps, rip currents, periodic circulation cells, and beautiful surge patterns in the coastal zone [1].

Several mechanisms for wave generation over a sloping beach have been proposed. Most of them involve the transfer of energy from incoming waves. On a sloping beach covered with a shallow water layer, a second-order nonlinear interaction occurs in the form of wave triads when two waves interact with each other, transferring energy to a third one. Gallagher [2] showed that the interaction between two incoming waves may result in resonant excitation of edge waves at the beat frequency. The idea of three-wave resonance in shallow water was successfully realized later (see [3–5]).

For small-scale edge waves, Guza and Davis [6] proposed a mechanism for the nonlinear interaction of edge waves with the incoming swell. Using the shallow-water approximation, they showed that a monochromatic harmonic wave train of frequency ω, normally incident and strongly reflected from the beach, is unstable with respect to subharmonic perturbations of standing edge waves of frequency $\omega/2$. Guza and Bowen [7] conducted a systematic study of the nonlinear mechanism of subharmonic resonance. An anomalously incident wave can form a resonant triad with two progressive edge waves traveling in opposite directions, generally with different frequencies.

Symonds *et al.* [8] proposed a breakpoint forcing mechanism. Assuming that the wave height in the surf zone depends only on the local depth, the modulated incident wave gives rise to excursions of the initial breakpoint on the temporal and spatial scales of wave modulation. The change in the position of the breakpoint in time leads to the generation of a group period wave (as well as its harmonics) and can be a considerable source of long-wave energy. Standing waves are found in the onshore direction from the breakpoint, while an outgoing progressive wave exists in the offshore direction. Although this model has considered only two-dimensional long waves, it is clear that a similar mechanism of breakpoint forcing can contribute significantly to the forcing of waves varying in the longshore direction. In particular, it provides an additional mechanism for the generation of free edge waves.

Another type of mechanism, connected to the action of external pressure, has also been proposed as a possibility. It has been observed that hurricanes and other similar events passing parallel to the coastline sometimes induce a resurgent wave motion. Munk *et al.* [9] suggested that this wave resurgence is primarily caused by the deviation of storm pressure from the normal atmospheric pressure distribution. Greenspan [10] demonstrated that large-scale edge waves can be excited by atmospheric forcing as a result of storm motion along a uniformly sloping coast. These waves are in "Greenspan resonance" with the moving pressure distribution, i.e., the velocity of the moving pressure anomaly is equal to the phase velocity of a particular edge wave mode. The typical period of such edge waves is related to the spatial extent of the storm zone and is of the order of a few hours. Some applications of this approach were discussed in other works [11–13].

Shrira *et al.* concluded, based on data from the SandyDuck '97 coastal experiment [14], that edge waves can be generated by wind [15]. They showed that the "maser" mechanism proposed by Longuet-Higgins [16] in the context of wave excitation on free water is effective under favorable conditions when interacting nonlinearly random wind-induced short waves generate a viscous shear stress on the water surface with stress phase variations related to edge waves. Thus, as in a maser, a single coherent harmonic edge wave is selected and amplified from a random field of linear edge waves. The higher the wind speed, the greater the magnitude of this effect.

The study of the "Munk-Greenspan" and "maser" mechanisms is based on the analysis of linearized fluid dynamics equations applied to a specially selected type of pressure ("external force"). The condition of constant pressure on the free surface was also not set in our study. The effect of wind is modeled by a pressure action with a fairly general distribution on the free surface. We use the Ptolemaic solutions; thus, the pressure is a harmonic function of time depending on two functions of spatial coordinates. But in contrast to other works [10,15], an exact solution of complete fluid dynamics equations is obtained. Note that it also describes unsteady waves with a vortex structure.

The theoretical study of edge waves was initiated by Stokes [17]. He proposed the simplest solution to the linearized problem over a homogeneous slope. The corresponding wave propagates along the coast in a positive or negative direction and decreases rapidly toward the open ocean. The dispersion relation for the Stokes edge wave is $\omega^2 = gk\sin\alpha$, where ω is the wave frequency, k is the wave number, g is the acceleration due to gravity, and α is the slope angle.

More precisely, discussing the work of Kelland [18], Stokes specified only a partial solution $\alpha = \pi/4$ (see [19, 20]). The Stokes edge wave is an example of a spatial wave when fluid particles move parallel to an inclined bank [21]. When $\alpha = \pi/2$, the solution of the Stokes edge wave corresponds to a linear potential gravity wave in deep water. In both waves, the fluid particles move in a circle. The Stokes solution is the lowest (main) mode among the infinite set of edge wave modes found on shores [22]. They are not considered in this study.

After their discovery, edge waves were for a long time regarded as a mathematical curiosity [19], but in the last third of the 20th century, their potential significance for physical processes in the coastal zone was emphasized (see [23–27] and references therein). Around the same time, a number of important analytical results relating to nonlinear Stokes edge wave theory were obtained. Whitham gave a description of a weakly nonlinear potential wave in the cubic approximation with respect to a small-wave steepness parameter [28]. A study of the properties of nonlinear progressive edge waves using the nonlinear Schrödinger equation was carried out by Yih [29].

232   *Analytical Fluid Dynamics in Lagrangian Variables*

Yih [30] and Mollo-Christensen [31] independently proposed an accurate description of a stationary edge wave traveling with constant velocity along the coast. Its analytical description is written in Lagrange variables and is similar to Gerstner's trochoidal wave [19, 32]. Constantin performed a detailed analysis of the wave [33] and specified explicit expressions for the trajectory coordinates of the fluid particles. Unlike all previously known examples, this strongly nonlinear wave exhibits vorticity. Similar to the Stokes edge wave, fluid particles in the Gerstner–Constantin solution move parallel to the sloping beach. Their trajectories are circles whose radii are the same for all particles along the shoreline (see Section 5.2).

This boundary wave is a traveling wave. Weber [34] in a special study noted that the usual superposition of two such waves of the same amplitude propagating in opposite directions does not give an exact solution. Nonlinear standing waves are also impossible.

13.1.   3D Exact Solution

We assume a constant plane beach and adopt the xyz coordinate system shown in Fig. 5.3 (Section 5.2), with the shoreline parallel to the x-axis and the steel sea in the region

$$b \le b_0, \quad 0 \le c \le (b_0 - b) \tan \alpha,$$

where $b_0 \le 0$ is a non-positive number, $0 < \alpha < \pi/2$, $\tan \alpha$ is the beach slope, and a, b, c are Lagrangian coordinates. We consider the fluid particle coordinates X, Y, Z as functions of these variables and time: $X = X(a, b, c, t); Y = Y(a, b, c, t); Z = Z(a, b, c, t)$.

The fluid dynamics equations of an ideal fluid in Lagrangian coordinates have the following form (see Eqs. (1.10) and (1.18)):

$$\frac{D(X, Y, Z)}{D(a, b, c)} = D_0(a, b, c), \tag{13.1}$$

$$X_{tt}X_a + (Y_{tt} + g \sin \alpha)\, Y_a + (Z_{tt} + g \cos \alpha)\, Z_a = -\frac{1}{\rho}\frac{\partial p}{\partial a}, \tag{13.2}$$

$$X_{tt}X_b + (Y_{tt} + g \sin \alpha)\, Y_b + (Z_{tt} + g \cos \alpha)\, Z_b = -\frac{1}{\rho}\frac{\partial p}{\partial b}, \tag{13.3}$$

$$X_{tt}X_c + (Y_{tt} + g \sin \alpha)Y_c + (Z_{tt} + g \cos \alpha)Z_c = -\frac{1}{\rho}\frac{\partial p}{\partial c}. \tag{13.4}$$

Here, $H = g(Y \sin \alpha + Z \cos \alpha)$. For a one-to-one correspondence between the coordinates of the particles and their labels ab and c, the function D_0 must be sign constant. Assume for certainty that $D_0 \geq 0$.

By cross differentiation of momentum Eqs. (13.2)–(13.4), we exclude the pressure and obtain three Cauchy invariants (see Section 1.3.3), namely,

$$\frac{D(X_t, X)}{D(b, c)} + \frac{D(Y_t, Y)}{D(b, c)} + \frac{D(Z_t, Z)}{D(b, c)} = S_1(a, b, c), \qquad (13.5)$$

$$\frac{D(X_t, X)}{D(c, a)} + \frac{D(Y_t, Y)}{D(c, a)} + \frac{D(Z_t, Z)}{D(c, a)} = S_2(a, b, c), \qquad (13.6)$$

and

$$\frac{D(X_t, X)}{D(a, b)} + \frac{D(Y_t, Y)}{D(a, b)} + \frac{D(Z_t, Z)}{D(a, b)} = S_3(a, b, c). \qquad (13.7)$$

Let us find nonlinear wave solutions of Eqs. (13.1), (13.5)–(13.7) satisfying the following conditions:

(a) no-flux condition at the bottom: $Z_t = 0$ for c = 0;
(b) absence of wave disturbances far away from the shore: X_t, Y_t, $Z_t \to 0$ for $b \to -\infty$, $c \to \infty$; this is a distinctive property of edge waves; and
(c) pressure is generally not constant on the free surface and is harmonic in time. Pressure variations correlate with coastal winds. The type of pressure distribution will be specified in the following.

We will look for solutions in the form

$$Z(a, b, c, t) = Z_*(c). \qquad (13.8)$$

This assumption means that the trajectories of motion of fluid particles are limited by the planes $Z_* = \text{const}$. For the flows described by Eq. (13.8), the normal velocity at the bottom is zero and the no-flux condition (a) is automatically satisfied.

With allowance for Eq. (13.8), Eqs. (13.5)–(13.7) can be rewritten as

$$\frac{D(X, Y)}{D(a, b)} = \frac{D_0(a, b, c)}{Z_*'(c)} = D_0^*(a, b, c), \qquad (13.9)$$

$$\frac{D(X_t, X)}{D(b,c)} + \frac{D(Y_t, Y)}{D(b,c)} = S_1(a,b,c), \qquad (13.10)$$

$$\frac{D(X_t, X)}{D(c,a)} + \frac{D(Y_t, Y)}{D(c,a)} = S_2(a,b,c), \qquad (13.11)$$

$$\frac{D(X_t, X)}{D(a,b)} + \frac{D(Y_t, Y)}{D(a,b)} = S_3(a,b,c). \qquad (13.12)$$

Let us introduce a new coordinate

$$q = b - c \qquad (13.13)$$

and assume that the functions X and Y depend only on a, q, and t, and $S_1 \equiv 0$. Then, Eq. (13.10) is automatically satisfied, and Eqs. (13.9), (13.11), and (13.12) have the form

$$\frac{D(X, Y)}{D(a,q)} = D_0^*(a,q), \qquad (13.14)$$

$$\frac{D(X_t, X)}{D(a,q)} + \frac{D(Y_t, Y)}{D(a,q)} = S_2(a,q), \qquad (13.15)$$

$$\frac{D(X_t, X)}{D(a,q)} + \frac{D(Y_t, Y)}{D(a,q)} = S_3(a,q). \qquad (13.16)$$

The left-hand sides of Eqs. (13.15) and (13.16) coincide, so the right-hand sides should also be equal, i.e., $S_2 = S_3$. We emphasize that neither the type of Cauchy invariants nor the right-hand sides of Eqs. (13.14)–(13.16) are initially specified; they will be calculated later based on the found solution of $X(a,q,t)$, $Y(a,q,t)$.

Thus, instead of three initial equations, only two equations can be solved, e.g., Eqs. (13.14) and (13.15). Formally, they coincide with the equations of two-dimensional fluid dynamics, where the variable q plays the role of the Cartesian Lagrangian coordinate b. However, the fluid motion is three-dimensional because of the dependence of the X, Y coordinates on the Lagrangian variable c. The transition to the case of planar 2D flows can be performed by assuming $c \equiv 0$ in Eq. (13.13).

Equations (13.14) and (13.15) have an exact solution (see Section 7.1):

$$\begin{aligned} W &= X + iY = G(\chi) \ \exp \ i\lambda t + F(\bar{\chi}) \exp i\mu t, \\ \chi &= a + i\,(b - c), \quad \text{and} \quad \bar{\chi} = a - i(b - c), \end{aligned} \qquad (13.17)$$

where G and F are analytical functions and λ and μ are real constants. The form of this expression coincides with the exact solution for 2D flows, but in our case, the solution depends on three Lagrangian coordinates. In each plane $Z_* = \text{const} > 0$, the fluid motion is different. However, the trajectories of fluid particles are qualitatively similar. Epicycloids are the trajectories for frequencies of the same sign and hypocycloids for frequencies of different signs. The solutions of Eq. (13.17) represent a particular case of Ptolemaic flows.

The pressure for the flows (13.17) is obtained from Eqs. (13.2)–(13.4) and is written as

$$
\begin{aligned}
\frac{p - p_0}{\rho} &= \frac{1}{2}\lambda^2 |G|^2 + \frac{1}{2}\mu^2 |F|^2 \\
&+ \mathrm{Re}\{e^{i(\lambda-\mu)t} \int (\lambda^2 G\bar{F}' + \mu^2 G'\bar{F})d\chi\} \\
&- g\mathrm{Im}\{Ge^{i\lambda t} + Fe^{i\mu t}\}\sin\alpha - gZ_* \cos\alpha.
\end{aligned}
\tag{13.18}
$$

The functions G, F and Z_* as well as the constants λ and μ will be specified in the following.

The Cauchy invariant S_2 is

$$
S_2 = 2(\lambda|G'|^2 - \mu|F'|^2).
$$

13.2. Gerstner–Constantin Solution

Let us first discuss the solution corresponding to a wave of constant shape propagating along the shore in the absence of wind (the pressure at the free surface is constant). Assume that $\lambda = 0$ and $G(\chi) = \chi$. In this case, the fluid particles rotate in a circle of radius $|F|$ with a certain frequency $\mu = \mu_*$. We are interested in the motion of a particle when the pressure on the free surface remains constant, i.e.,

$$
p\big|_{c=(b_0-b)\tan\alpha} = \text{const.}
$$

The pressure is time-independent when $F(\bar{\chi}) = \frac{i}{k}\exp ik\bar{\chi}$; $\mu_*^2 = gk\sin\alpha$, where k is the wave number and the second expression is the dispersion relation for the wave. Taking into account the expressions for G and F, the solution (13.17) can be written in real form as

$$X = a - \frac{1}{k} e^{kq} \sin(ka + \mu_* t), \quad q = b - c,$$

$$Y = q + \frac{1}{k} e^{kq} \cos(ka + \mu_* t). \tag{13.19}$$

Assuming $c \equiv 0$, Eq. (13.19) describes the classical Gerstner wave.

Yih [30] and Mollo-Christensen [31] found that a simple transformation of the Gerstner solution gives an exact solution to the edge wave problem. However, a more complete and useful analysis of this aspect of the Gerstner solution was developed by Constantin [33], who not only gave a clear realization of the solution but also demonstrated its mathematical correctness.

Substituting Eq. (13.19) into the expression for pressure (13.18) yields

$$p - p_0 = \frac{\rho g \sin \alpha}{2k} e^{2k(b-c)} - \rho g (b - c) \sin \alpha - \rho g Z_*(c) \cos \alpha.$$

The function $Z_*(c)$ determines the relation between the Eulerian and Lagrangian vertical coordinates, and hence governs the pressure distribution as well. Following Constantin [33], we will choose this function as in Eq. (5.11). The values $c = 0$ and $Z_* = 0$ correspond to the bottom and shoreline positions, respectively; $0 < c < (b_0 - b) \tan \alpha$ (see Fig. 5.3).

With allowance for Eq. (5.11), the expression for the pressure field has the form

$$\frac{p - p_0}{\rho g} = \frac{\sin \alpha}{2k} \left[e^{2k(b-c)} \left(1 - e^{2k(b_0 - b - c \cot \alpha)} \right) - e^{2kb_0} \right]$$

$$- (b + c \cdot \cot \alpha) \sin \alpha.$$

It follows from this expression that the pressure on the free surface is constant, $p_0 = p_{\text{atm}} - \rho g (b_0 + \frac{1}{2k}) \sin \alpha$, where p_{atm} is the atmospheric pressure.

Thus, Eq. (13.19) yields the solution in the form of an edge trochoidal wave on a sloping beach. It propagates to the left along the shoreline at a constant velocity μ_*/k. The amplitude of the wave has its maximum in the vicinity of the shore and decreases exponentially seaward (with increasing c and b).

Despite its relatively simple form, Eq. (13.19) represents an exact solution to the extremely complicated wave problem. It reveals the most important features of the Lagrangian description of the flow. The coordinates a, b, c are the labels of the fluid particles. Here, they do not coincide with the initial positions of the particles, but the condition of one-to-one correspondence between the coordinates X, Y, Z and their labels a, b, c is fulfilled. Indeed,

$$D_0 = Z'_* D_0^* = (1 + \tan \alpha)(1 - e^{-2kc(1+\cot \alpha)})[1 - e^{2k(b-c)}].$$

That is why the condition $D_0 \geq 0$ is fulfilled within the flow at $b \leq b_0$ and $c \geq 0$. For $b = b_0 = 0$ and $c = 0$, when $D_0 = 0$, Eq. (13.19) describes the limiting edge waves with a cycloidal profile when the wave crests sharpen.

Shoreward run-up patterns are formed by standing waves on a sandy shallow beach. The Gerstner–Constantin edge wave is a unidirectional propagating wave, so it cannot form steady shoreward run-up patterns. However, one can specify a generalization of Eq. (13.19) that is free from this shortcoming.

13.3.   Gerstner-Like Waves

Now, let the motion of the fluid along the sloping beach be described by the following relations [35]:

$$W = G(\chi) + F(\bar{\chi})e^{i\mu t} \tag{13.20}$$

$$Z = Z_* = c + c \cdot \tan \alpha - \frac{\tan \alpha}{2k} e^{2kb_0}[1 - e^{-2kc(1+\cot \alpha)}]. \tag{13.2}$$

From the family of exact solutions (13.17), we choose one with $\lambda =$ where μ is a free parameter, generally different from the frequ μ_* of the Stokes edge wave, and the form of the function $Z(c)$ responds to other work [33]. Each of the fluid particles moves i plane $Z_* = $ const in a circle with frequency μ and radius $|F'|$ position of the center of each circle is determined by the type o tion G. For clarity of the description of flows (13.20) and (13.2

238   *Analytical Fluid Dynamics in Lagrangian Variables*

required that

$$D_0^* = \frac{J(X,Y)}{J(a,q)} = \frac{J(W,\bar{W})}{J(\chi,\bar{\chi})} = |G'|^2 - |F'|^2 \geq 0. \qquad (13.22)$$

The zero value of the Jacobian can be obtained only on the free surface (at the cusps): This rule is well illustrated by the Gerstner wave. Condition (13.22) is not too strict, so the functions G and F can be quite arbitrary. As a consequence, Eqs. (13.20) and (13.21) describe a complex non-stationary wave oscillation of the free surface, with the velocity field defined by two functions.

We will search for edge waves, that is, waves that are periodic in x (along the shore) and exponentially decreasing as we move away from the shore (c and $-b$ increasing). Assume that

$$G(\chi) = \chi + \sum_{n=1}^{N} \alpha_n e^{-ikn\chi}, \quad F(\bar{\chi}) = \sum_{m=1}^{M} \beta_m e^{ikm\bar{\chi}}, \qquad (13.23)$$

here N and M are the positive integers greater than unity, and
and β_m are constants of the length dimension. Equation (13.23)
res that condition (b) is satisfied. To fulfill condition (13.22), it
icient to require that

$$\sum_{n=1}^{N} kn \, |\alpha_n| e^{knb_0} + \sum_{m=1}^{M} km \, |\beta_m| e^{kmb_0} \leq 1.$$

ness of the function $G(\chi)$ follows from this inequality.
13.20) and (13.21) generalize the Gerstner–Constantin
describe a more complex class of wave motions, where
move in circles as in the Gerstner wave. Therefore, the
Eqs. (13.20) and (13.21) can be called Gerstner-like

eld within the fluid is determined by the formula

$$\mathrm{Re}[(\mu^2 \int G' \bar{F} d\chi - ig\bar{F})e^{-i\mu t}]$$

$$\left(\frac{k \sin \alpha}{2} |F|^2 - \mathrm{Im}G - Z_* \cos \alpha \right) g.$$

ribution on the free surface, we assume
xpression. It is a function of two spatial

coordinates, a and b, and time. The effect of wind is reduced to the action of pressure on a free surface that is non-uniform in spatial coordinates and harmonic in time. The wave amplitude growth is caused by the action of external pressure, similar to the case described in Greenspan's paper [10].

An important property of the considered waves is their vorticity. Since the motion of fluid particles occurs in the z-plane, only the z-component of vorticity differs from zero. It is defined by the following expression:

$$\Omega_z = \frac{D(Y_t, Y, Z)}{D(X, Y, Z)} - \frac{D(X, X_t, Z)}{D(X, Y, Z)} = -\frac{2\mu\,|F'|^2}{|G'|^2 - |F'|^2}.$$

For the flows (13.20), (13.21), and (13.23), the vorticity is written as

$$\Omega_z = -\frac{2\mu k^2 \left|\sum m\beta_m e^{ikm\bar{\chi}}\right|^2}{\left|1 - ik\sum n\alpha_n e^{-ikn\chi}\right|^2 - k^2\left|\sum m\beta_m e^{ikm\bar{\chi}}\right|^2}, \qquad (13.24)$$

where the summation is performed within the limits specified in Eq. (13.23). In the case

$$\alpha_n = 0, \quad n = 1, 2, \ldots, N; \quad \mu = \mu_*; \quad \beta_1 = 1/k; \quad \beta_m = 0,$$
$$m = 2, 3, \ldots, M,$$

according to the Gerstner–Constantin solution, the vorticity is [33]

$$(\Omega_z)_{GC} = -\frac{2\mu_* e^{2k(b-c)}}{1 - e^{2k(b-c)}}.$$

Obviously, in the general case, expression (13.24) specifies a very complex vorticity distribution depending on three Lagrangian variables. Note that, despite the fact that the flow given by Eqs. (13.20) and (13.21) is not two-dimensional, the vorticity of each individual fluid particle is conserved.

The form of the free surface is given parametrically:

$$X(a, b) = \mathrm{Re}\left[G(\chi_0) + F(\overline{\chi_0})\,e^{i\mu t}\right]; \quad \chi_0 = a + it[\gamma b_0 + (1 + \gamma)b], \tag{13.25}$$

$$Y(a, b) = \mathrm{Im}\left[G(\chi_0) + F(\overline{\chi_0})\,e^{i\mu t}\right], \quad \gamma = \tan\alpha, \tag{13.26}$$

$$Z(b) = \gamma(1+\gamma)(b_0 - b) - \frac{\gamma}{2k}e^{2kb_0}\{1 - \exp[-2k(1+\gamma)(b_0 - b)]\},$$

$$(13.27)$$

where the parameters are Lagrangian variables $-\infty < a < \infty$; $b \le b_0$. The motion of a particle on the free surface can conveniently be represented as follows. According to Eq. (13.27), the value of $Z = Z(b_1)$ corresponds to a particle with coordinates $a = a_1$ and $b = b_1$ defining the plane of motion of the parcel (see Fig. 1.1). Equations (13.25) and (13.26), written for $\chi_{01} = a_1 + i[\gamma b_0 + (1+\gamma)b_1]$, serve as parametric expressions for the particle trajectory, which is a circle centered at the point $G(\chi_{01})$ with radius $|F(\overline{\chi_{01}})|$. Particles with $a \ne a_1$ and $b = b_1$ move in the same plane and also in a circle, but the radius and phase of their rotation will be different. This distinguishes Gerstner-like waves from the Gerstner–Constantin solution. Particles with $a = a_1$ and $b \ne b_1$ rotate in phase in parallel planes over circles whose radii decrease exponentially with an increase in distance from the shore. Thus, the motion of the free surface resembles a flag waving in the wind (this qualitative result will be confirmed further by numerical analysis).

The amplitude of the edge wave is maximum near the shore and decreases seaward. This corresponds to the fact that the free surface near the shore rises relative to the plane

$$Z + (Y - b_0)\tan\alpha = 0, \qquad (13.28)$$

coinciding with the calm water level (Lagrangian shape of the free surface), and asymptotically approaches another horizontal plane located below with the water level being "at infinity" toward the sea. Let us denote the distance between these parallel planes as h and measure the amplitude of the wave A from the lower of the planes. In this case, the distance of an arbitrary point on the free surface to the plane (13.28) is equal to

$$A - h = (Y - b_0)\sin\alpha + Z\cos\alpha, \qquad (13.29)$$

with $A \to 0$ for $b \to -\infty$. Substituting Eqs. (13.26), (13.27), and (13.23) into Eq. (13.29) yields

$$A(\sin\alpha)^{-1} = \mathrm{Im}\left[\sum_{n=1}^{N}\alpha_n e^{-ikn\chi_{01}} + e^{i\mu t}\sum_{m=1}^{M}\beta_m e^{ikm\overline{\chi_{01}}}\right]$$

$$+ \frac{1}{2k}e^{2k[(1+\gamma)b-\gamma b_0]} \quad \text{and} \quad h = \frac{\sin\alpha}{2k}e^{2kb_0}.$$

The amplitude $A = A(a, b, t)$ is a function of two spatial coordinates and time. As follows from its representation, the unsteady edge wave varies periodically along the horizontal coordinate a (with period $2\pi/k$) and exponentially decreases away from the shore. Harmonic oscillations of fluid particles and oscillations of the free surface points occur with frequency μ.

The evolution of fluid-shore boundary is described by Eqs. (13.20) and (13.23) for $b = b_0$ and $c = 0$. The wave profile is located in the $Z = 0$ plane and is composed of fluid particles rotating in a circle of radius $|F(a - ib_0)|$. The value of this radius varies over the wave period. The circle centers lie on the curve, which is defined parametrically, $X = \operatorname{Re} G(a + ib_0)$; $Y = \operatorname{Im} G(a + ib_0)$ and can be arbitrary (see Eq. (13.22)).

The drift current along the shoreline is absent due to the rotation of fluid particles along the circles, i.e., there is no mass transfer in the direction of wave propagation. For the considered unsteady oscillations, it is also possible that at certain points (nodes) the radius of oscillation of fluid particles turns to zero (characteristic of standing waves). Such oscillations of the fluid boundary are illustrated in Fig. 13.1. For simplicity, we have chosen the following values of variables in Eqs. (13.23), (13.25), and (13.26):

$$b_0 = 0; \quad k = 1; \quad \leq a \leq 2\pi; \quad \alpha_1 = 0; \quad \beta_1 = \beta_2 = 0.35,$$

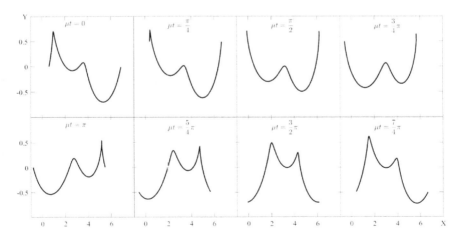

Fig. 13.1. Evolution of the wave profile denoting the coastline.

242   *Analytical Fluid Dynamics in Lagrangian Variables*

all the rest (α_n and β_m) are equal to zero. Since

$$|F| = \sqrt{\beta_1^2 + \beta_2^2 + 2\beta_1\beta_2 \cos a},$$

the point corresponding to $a = \pi = X$ is the node and the standing point. At the end of the depicted spatial period ($a \in [0, 2\pi]$), the displacement from the zero level is maximum. In its middle, the fluid particles have a smaller displacement. Thus, the considered wave forms inhomogeneous "standing" splashes along the shore. For $\beta_1 \neq \beta_2$, the nodes are absent.

The considered example corresponds to particular types of the functions G and F. Another representation of the solution makes it possible to extend the range of the described phenomena. Thus, a specific choice of the function G takes into account the curvilinearity of the beach boundary, while varying the type of the function F leads to a more complex evolution of the edge wave profile.

Let us assume that $b_0 = 0$ for the time $t = 0$. The wave profile $Y^0(X^0)$, denoting the coastline, is parametrically given by the following expressions:

$$X^0(a) = \mathrm{Re}[G(a) + F(a)]; \quad Y^0(a) = \mathrm{Im}\,[G(a) + F(a)],$$

where the functions G and F are defined by Eq. (13.23), which represents their Fourier series expansions. Obviously, these functions may be arbitrary. Similarly, the wave profile may take the form of an arbitrary curve without intersections (the condition of a non-zero Jacobian D_0^* is guaranteed by Eq. (13.22)). Thus, our model is valid for the evolution of coastal splashes of different initial forms. Therefore, the considered solutions give a qualitative insight into the different fluid motion regimes observed in the coastal zone.

The obtained solutions can be used to explain the dynamics of foam, entrained air bubbles, or unattached weeds. Particles of these "impurities" have circular trajectories in a given velocity field, similar to a passive scalar. However, a significant complication of the fluid dynamics model is required to describe the formation of coastal structures as a result of sediment movement.

References

[1] Komar, P. (1998). *Beach Processes and Sedimentation*, 2nd ed. (New York, Prentice-Hall).

[2] Gallagher, B. (1971). Generation of surf beat by nonlinear wave interactions. *J. Fluid Mech.*, 49, pp. 1–20.

[3] Freilich, M.H. and Guza, R.T. (1984). Nonlinear effects on shoaling surface gravity waves. *Phil. Trans. R. Soc. Lond. A*, 311(1515), pp. 1–41.

[4] Kirby, J.T., Putrevu, U. and Özkan Haller, H.T. (1998). *Evolution equations for edge waves and shear waves on longshore uniform beaches*. In: Proc. 26th International Conference on Coastal Engineering, pp. 203–216.

[5] Sheremet, A., Davis, J.R., Tian, M., Hanson, J.L. and Hathaway, K.K. (2016). TRIADS: A phase-resolving model for nonlinear shoaling of directional wave spectra. *Ocean Model.*, 99, pp. 60–74.

[6] Guza, R.T. and Davis, R.E. (1974). Excitation of edge waves by waves incident on a beach. *J. Geophys. Res.*, 79(9), pp. 1285–1291.

[7] Guza, R.T. and Bowen, A.J. (1976). Finite amplitude edge waves. *J. Marine Res.*, 34, pp. 269–293.

[8] Symonds, G., Huntley, D.A. and Bowen, A.J. (1982). Two-dimensional surf beat: Long wave generation by a time-varying breakpoint. *J. Geophys. Res.*, 87(C1), pp. 492–498.

[9] Munk, W., Snodgrass, F. and Carrier, G. (1956). Edge waves on the continental shelf. *Science*, 123, pp. 127–132.

[10] Greenspan, H.P. (1956). The generation on edge waves by moving pressure distributions. *J. Fluid Mech.*, 1, pp. 574–592.

[11] Viera, F. and Buchwald, V.T. (1983). The response of the East Australian continental shelf to a travelling pressure disturbance. *Geophys. Astrophys. Fluid Dyn.*, 19, 249.

[12] Seo, S.N. and Liu, P.L-F. (2014). Edge waves generated by atmospheric pressure disturbances moving along a shoreline on a sloping beach. *Coastal Eng.*, 85, pp. 43–59.

[13] Seo, S.N. (2015). Edge waves generated by external forces, *Proc. Eng.*, 116, pp. 390–397.

[14] Birkemeier, W., Long, A. and Hathaway, K.K. (1997). DELILAH, DUCK94 & SandyDuck: Three nearshore field experiments. In: *Proceedings of the 25th International Conference on Coastal Engineering* (Orlando FL ASCE).

[15] Shrira, V.I., Sheremet, A., Troitskaya, Y.I. and Soustova, I.A. (2022). Can edge waves be generated by wind? *J. Fluid Mech.*, 934, A16–36.

[16] Longuet-Higgins, M.S. (1969). A nonlinear mechanism for the generation of sea waves. *Proc. R. Soc. Lond.* A, 311(1506), pp. 371–389.

[17] Stokes, G.G. (1846). Report on recent researches in hydrodynamics. *Rep. 16th Brit. Assoc. Adv. Sci.*, pp. 1-20; see also: (1880) *Papers*, Vol. 1, pp. 157–187 (Cambridge University Press, Cambridge).

[18] Kelland, P. (1844). On the theory of waves. Part 2. *Trans. R. Soc. Edin.*, 15, pp. 101–144.

[19] Lamb, H. (1932). *Hydrodynamics*, 6th ed. (Cambridge University Press, Cambridge).

[20] Johnson, R.S. (2007). Edge waves: Theories past and present. *Phil. Trans. R. Soc. A*, 365, pp. 2359–2376.

[21] Stuhlmeier, R. (2015). Particle paths in Stokes' edge wave. *J. Nonl. Math. Phys.*, 22(4), pp. 507–515.

[22] Ursell, F. (1952). Edge waves on a sloping beach. *Proc. R. Soc. A*, 214, pp. 79–97.

[23] Leblond, P.H. and Mysak, L. (1978). *Waves in the ocean*, 1st ed. (*Elsev. Ocean. Series*, 20).

[24] Bowen, A.J. and Inman, D.L. (1971). Edge waves and crescentic bars. *J. Geoph. Res.*, 76, 30, pp. 8662–8671.

[25] Guza, R.T. and Inman, D.L. (1975). Edge waves and beach cusps. *J. Geoph. Res.*, 80(20), pp. 2997–3012.

[26] Bowen, A.J. and Guza, R.T. (1978). Edge waves and surf beat. *J. Geophys. Res.*, 83(C4), pp. 1913–1920.

[27] Holman, R.A. and Bowen, A.J. (1982). Bars, bums, and holes: Models for the generation of complex beach topography. *Geoph. Res.*, 87(C1), pp. 457–468.

[28] Whitham, G.B. (1976). Nonlinear effects in edge waves. *J. Fluid Mech.*, 74, pp. 353–368.

[29] Yeh, H. (1985). Nonlinear progressive edge waves: Their instability and evolution. *J. Fluid Mech.*, 152, pp. 479–499.

[30] Yih, C.-S. (1982). Note on edge waves in a stratified fluid. *J. Fluid Mech.*, 24, pp. 765–767.

[31] Mollo-Christensen, E. (1982). Allowable discontinuities in Gertner wave. *Phys. Fluids*, 2, pp. 586–587.

[32] Gerstner, F. (1809). Theorie der Wellen samt einer daraus abgeleiteten Theorie der Deichprofile. *Ann. Phys.*, 2, pp. 412–455.

[33] Constantin, A. (2001). Edge waves along a sloping beach. *J. Phys. A: Maths. Gen.*, 34, pp. 9723–9731.

[34] Weber, J.E.H. (2012). A note on trapped Gerstner waves. *J. Geophys. Res.*, 117, C03048.

[35] Abrashkin, A.A. (2022). Unsteady edge waves generated by time-harmonic pressure: Exact solutions. *J. Phys. A: Math. Theor.*, 55, 415701.

Part V

Viscous Fluid Flows

Chapter 14

Basic Equations and Examples

14.1. Navier–Stokes Equations

So far, we have focused our efforts on the ideal fluid. How will the equations in the Lagrange variables change with allowance for viscosity? The continuity equation will obviously retain its form. It is convenient to write it as follows (we are still limited to analyzing the motion of an incompressible fluid):

$$\frac{\partial}{\partial t}[X_1, X_2, X_3] = \frac{\partial}{\partial t}D_0 = 0. \tag{14.1}$$

Here, the following notations were used:

$$X_1 = X(a, b, c, t), \quad X_2 = Y(a, b, c, t), \quad \text{and} \quad X_3 = Z(a, b, c, t),$$

Square brackets denote the Jacobian in variables a, b and c. The Jacobian does not change its magnitude with cyclic permutation of the functions X_1, X_2, and X_3 and changes sign with acyclic permutation.

The Navier–Stokes equation in Eulerian variables can be written as follows:

$$\frac{dV_i}{dt} = -\frac{1}{\rho}\frac{\partial p}{\partial X_i} + \nu\Delta_{\vec{X}}\,V_i, \tag{14.2}$$

where V_i refers to the velocity components and ν is the viscosity. Since

$$\frac{dV_i}{dt} = X_{itt},$$

it is necessary to find a Lagrangian representation for the pressure and viscous forces to write the equation in Lagrangian form.

The derivative of the function $f(X_i)$ is expressed in terms of derivatives with respect to Lagrangian variables as follows:

$$\frac{\partial f}{\partial X_i} = \frac{1}{D_0}[X_j, X_k, f]. \tag{14.3}$$

Here, the combination of indices (i, j, and k) always means cyclic permutation $(1, 2, 3)$. This expression is obtained in a manner similar to Eq. (1.47). If pressure is chosen as the function f, Eq. (14.2) determines (with allowance for the factor $-1/\rho$) the expression for the pressure forces in the Navier–Stokes Eq. (14.2).

To allow for viscosity, i.e., determine the formula for the Laplace operator in Lagrangian variables, one should use Eq. (14.3) twice:

$$\Delta_{\vec{X}} f = \frac{\partial}{\partial X_\alpha}\frac{\partial f}{\partial X_\alpha} = D_0^{-1}\{[X_2, X_3, D_0^{-1}[X_2, X_3, f]]$$
$$+ [X_3, X_1, D_0^{-1}[X_3, X_1, f]] + [X_1, X_2, D_0^{-1}[X_1, X_2, f]]\}. \tag{14.4}$$

Using Eqs. (14.3) and (14.4), we obtain

$$\frac{\partial^2 X_i}{\partial t^2} = -(\rho D_0)^{-1}[X_j, X_k, p] + \nu D_0^{-1}[X_2, X_3, D_0^{-1}[X_2, X_3, \partial X_i/\partial t]$$
$$+ [X_3, X_1, D_0^{-1}[X_3, X_1, \partial X_i/\partial t]] \tag{14.5}$$
$$+ [X_1, X_2, D_0^{-1}[X_1, X_2, \partial X_i/\partial t].$$

Equations (14.1) and (14.5) with respect to the unknowns $X_i(\vec{a}, t)$, $i = 1, 2, 3$ and $p(\vec{a}, t)$ form a complete system of equations of dynamics of an incompressible viscous fluid in Lagrangian variables. While in the Eulerian description the terms nonlinear with respect to V_i are contained in the expression for the acceleration dV_i/dt (see Eq. (14.2)), and the viscosity forces are described by linear expressions, the case is just the opposite in the Lagrangian description. Inertial interactions between inhomogeneities of the velocity field, described by the expression $(\vec{V}\nabla)V_i$, are relative in nature and are eliminated in the accompanying (Lagrangian) reference frame. However, in it, the viscosity forces become nonlinear; moreover, they are described by nonlinear expressions of the fifth degree with respect to the variables X_i. The expression for pressure also becomes much

Basic Equations and Examples 249

more complicated. Now, it is also nonlinear. Note that the Lagrangian approach significantly complicates the problem of pressure calculation from a known velocity field.

In the case of the "Eulerian" Navier–Stokes Eq. (14.2), the measure of nonlinearity is the Reynolds number Re, i.e., the ratio of typical values of inertia forces $(\vec{V}\nabla)V_i$ and viscosity forces $\nu\Delta V_i$. At sufficiently high Re, inertial interactions turn out to be very strong and are dominant. In the Lagrangian description, the measure of nonlinearity is already the ratio of typical values of viscosity forces to typical acceleration, i.e., 1/Re. At high Re, viscous interactions turn out to be very weak.

The form of Eqs. (14.1) and (14.5) is fairly complicated. It is quite natural to simplify them by choosing the initial positions of the particles as Lagrangian coordinates. As a result, we have

$$[X_1, X_2, X_3] = D_0 = 1,$$

$$\begin{aligned}
\frac{\partial^2 X_i}{\partial t^2} =\ & -\rho^{-1}[X_j, X_k, p] + \nu[X_2, X_3, [X_2, X_3, \partial X_i/\partial t]] \\
& + [X_3, X_1, [X_3, X_1, \partial X_i/\partial t]] \\
& + [X_1, X_2, [X_1, X_2, \partial X_i/\partial t]].
\end{aligned} \tag{14.6}$$

The derivation of these equations was apparently published for the first time only in the middle of the 20th century by Gerber [1] Somewhat later, the same result was independently obtained by A.S. Monin [2] (see also the monograph [3]).

Here is another type of equation for a two-dimensional plane-parallel flow. Let the motion occur only in the plane $Z = c = \text{const}$, and X_1 and X_2 do not depend on c. Then, using the notation

$$\frac{\partial(A, B)}{\partial(a, b)} = [A, B]$$

for Jacobians on (a, b), we write Eq. (14.6) in the following form:

$$[X_1, X_2] = 1,$$

$$\frac{\partial^2 X_1}{\partial t^2} = -\rho^{-1}[p, X_2] + \nu\left\{\left[X_1, \left[X_1, \frac{\partial X_1}{\partial t}\right]\right] + \left[X_2, \left[X_2, \frac{\partial X_1}{\partial t}\right]\right]\right\},$$

$$\frac{\partial^2 X_2}{\partial t^2} = -\rho^{-1}[X_1, p] + \nu \left\{ \left[X_1, \left[X_1, \frac{\partial X_2}{\partial t}\right]\right] + \left[X_2, \left[X_2, \frac{\partial X_2}{\partial t}\right]\right] \right\}.$$
$$(14.7)$$

14.2. Equations for Cauchy Invariants

In the previous paragraph, the Lagrangian momentum equations of a viscous fluid were obtained by proceeding to differentiation with respect to Lagrangian variables in the Navier–Stokes equation itself. In this section, we derive another form of these equations, where Cauchy invariants act as unknown functions [4]. In the case of a viscous fluid, the quantities S_1, S_2, and S_3 (i.e., circulation over elementary Lagrangian sites) already depend on time, but we still call them invariants, as for an ideal fluid.

We now write the Navier–Stokes Eq. (14.2) using the well-known identity for the Laplacian of the solenoid vector:

$$\vec{R}_{tt} = -\frac{1}{\rho} \nabla p - \nu \operatorname{rot}_{\vec{X}}(\operatorname{rot}_{\vec{X}} \vec{R}_t).$$

Multiply this equation on the left by the matrix $\hat{R}^T$ (see Section 1.2). Then, we obtain

$$\hat{R}^T \vec{R}_{tt} = -\frac{1}{\rho} \nabla_{\vec{a}} p - \nu \hat{R}^T \operatorname{rot}_{\vec{X}} (\operatorname{rot}_{\vec{X}} \vec{R}_t).$$
$$(14.8)$$

Recall that the subscripts $\vec{a}$ and $\vec{X}$ of the sign of the differential operation mean that the differentiation is performed with respect to Lagrangian or Euler variables, respectively.

To exclude the pressure, we apply to Eq. (14.8) the operation of taking the curl with respect to Lagrangian variables:

$$\operatorname{rot}_{\vec{a}} \left(\hat{R}^T \vec{R}_{tt} \right) = -\nu \operatorname{rot}_{\vec{a}} \left(\hat{R}^T \operatorname{rot}_{\vec{X}} (\operatorname{rot}_{\vec{X}} \vec{R}_t) \right).$$
$$(14.9)$$

So far, we are repeating the chain of transformations carried out in the first chapter for an ideal fluid. If we assume $\nu = 0$, then after integrating the left-hand side in time, we obtain the system (1.23). In particular, it follows from this that the left-hand side of Eq. (14.9) is equal to the time derivative of the vector of Cauchy invariants, i.e.,

$$\vec{S}_t = \text{rot}_{\vec{a}}(\hat{R}^T \vec{R}_{tt}). \tag{14.10}$$

To allow for viscosity, we should proceed to differentiation by Lagrangian coordinates on the right-hand side of Eq. (14.9). Let us use the formula [5]:

$$\text{rot}_{\vec{X}}\, \vec{A} = \frac{\hat{R}}{D}\, \text{rot}_{\vec{a}}(\hat{R}^T \vec{A}), \tag{14.11}$$

where $\vec{A}$ is an arbitrary vector and $D = \det \hat{R}$. The validity of this formula can be verified by a special example if we assume $\vec{A} = \vec{R}_t$. Then, the left-hand side of Eq. (14.11) is equal to the vorticity. The expression for the curl on the right-hand side, as can be verified by direct calculation, can be written as follows:

$$\text{rot}_{\vec{a}}(\hat{R}^T \vec{R}_t) = \vec{S}. \tag{14.12}$$

With allowance for this relation, we conclude that in the example under consideration, Eq. (14.11) turns into the relation for vorticity (1.39).

To write Eq. (14.9) in Lagrangian form, we employ Eq. (14.11) twice. As a result, we obtain the following equation:

$$\text{rot}_{\vec{a}}(\hat{R}^T \vec{R}_{tt}) = -\nu \,\text{rot}_{\vec{a}}(D^{-1}\hat{R}^T \hat{R} \text{rot}_{\vec{a}}(D^{-1}\hat{R}^T \hat{R} \,\text{rot}_{\vec{a}}(\hat{R}^T \vec{R}_t))). \tag{14.13}$$

Taking into account relations (14.10) and (14.12) and introducing a new matrix

$$\hat{g} = \hat{R}^T \hat{R}; \quad \hat{g} = g_{ij}; \quad i, j = 1, 2, 3;$$
$$g_{11} = X_a^2 + Y_a^2 + Z_a^2; \quad g_{12} = g_{21} = X_a X_b + Y_a Y_b + Z_a Z_b; \tag{14.14}$$
$$g_{22} = X_b^2 + Y_b^2 + Z_b^2; \quad g_{13} = g_{31} = X_a X_c + Y_a Y_c + Z_a Z_c;$$
$$g_{33} = X_c^2 + Y_c^2 + Z_c^2; \quad g_{23} = g_{32} = X_b X_c + Y_b Y_c + Z_b Z_c,$$

we obtain the equation for the Cauchy invariants

$$\vec{S}_t = -\nu \,\text{rot}_{\vec{a}}\left(D^{-1}\hat{g}\,\text{rot}_{\vec{a}}(D^{-1}\hat{g}\,\vec{S})\right). \tag{14.15}$$

252 *Analytical Fluid Dynamics in Lagrangian Variables*

The determinant D coincides with the value of the right-hand side of the continuity equation. If the initial positions of the particles are chosen as Lagrange coordinates, then it is equal to one, and the form of Eq. (14.15) is simplified:

$$\vec{S}_t = -\nu \, \mathrm{rot}_{\bar{a}}(\hat{g} \, \mathrm{rot}_{\bar{a}}(\hat{g} \, \vec{S})). \qquad (14.16)$$

Equations (14.12) and (14.15) are exactly the required equations for Cauchy invariants. It is important to note that Eq. (14.12) in this system of equations coincides with Eq. (1.23) for an ideal fluid. There was a sort of separation of the inertial processes described by Eq. (14.12) and viscous processes described by Eq. (14.15).

In the case of a two-dimensional flow, Eq. (14.15) is somewhat simplified. The Cauchy invariant vector has only one component directed along the c-axis (so that $\vec{S} = \vec{S}_3$). In addition, we take into account that

$$g_{ij}\vec{S} = \vec{S}_3,$$

and Eq. (14.15) takes the following form (index 3 is omitted):

$$S_t \, \vec{c}^{\,0} = -\nu \, \mathrm{rot}_{\bar{a}} \left(D^{-1}\hat{g} \left(\vec{a}^{\,0} \frac{\partial}{\partial b} \frac{S}{D} - \vec{b}^{\,0} \frac{\partial}{\partial a} \frac{S}{D} \right) \right). \qquad (14.17)$$

In scalar form, the equation will be written as follows:

$$
\begin{aligned}
S_t = - \nu \Bigg\{ & \left(\frac{g_{11}}{D} \left(\frac{S}{D} \right)_b - \frac{g_{12}}{D} \left(\frac{S}{D} \right)_a \right)_b - \left(\frac{g_{21}}{D} \left(\frac{S}{D} \right)_b \right. \\
& \left. - \frac{g_{22}}{D} \left(\frac{S}{D} \right)_a \right)_a \Bigg\};
\end{aligned} \qquad (14.18)
$$

$$g_{11} = X_a^2 + Y_a^2; \quad g_{12} = g_{21} = X_a X_b + Y_a Y_b;$$

$$g_{22} = X_b^2 + Y_b^2; \quad D = X_a Y_b - X_b Y_a.$$

Here, the indices at the bottom mean differentiation by the corresponding Lagrange variable. We emphasize that due to the continuity equation, the value of D should not depend on time. As in the three-dimensional case, Eqs. (14.17) and (14.18) are simplified if we assume $D = 1$.

Equations (14.15) and (14.17) are equivalent to system (14.5). However, they do not contain pressure. If they are resolved, the found

Basic Equations and Examples   253

solution should be substituted into initial Eq. (14.8) in order to find the pressure.

Both forms of viscous equations are unusual for fluid mechanics, so it makes sense to test them with some specific examples. As such, we will choose the known exact solutions obtained earlier in Euler variables.

14.3.   Exact Solutions

Let us start with the simplest cases of rectilinear motion of fluid particles.

14.3.1.   *Flow in the channel and the Couette flow*

The plane shear flow in Lagrangian variables is given by the relations

$$X = a + U(b)t \quad \text{and} \quad Y = b, \tag{14.19}$$

where $U(b)$ determines the flow velocity profile. This type of flow corresponds to a two-dimensional Jacobi matrix

$$\hat{R} = \begin{pmatrix} 1 & U't \\ 0 & 1 \end{pmatrix}.$$

Its determinant D is equal to one, and the metric tensor $\hat{g}$ related to it has the form

$$\hat{g} = \begin{pmatrix} 1 & U't \\ U't & 1 + U'^2t^2 \end{pmatrix}. \tag{14.20}$$

We should also calculate the components of the Cauchy invariant vector. For a two-dimensional flow, it will have only one component S_3. As follows from Eq. (14.12), its value is equal to

$$S_3 = -U'(b). \tag{14.21}$$

The value of the invariant coincides with the vorticity of the flow. By substituting Eqs. (14.20) and (14.21) into Eq. (14.18), we obtain a condition for the shape of the shear flow profile:

$$U''' = 0.$$

The general solution to this condition is written as follows:

$$U = \alpha b^2 + \beta b + \gamma. \tag{14.22}$$

The constant α is related to the pressure distribution in the fluid. Substituting the solution Eq. (14.19) into the system (14.7), we obtain that

$$\alpha = \frac{1}{2\nu\rho}\frac{dp}{da} = \text{const},$$

i.e., Eq. (14.19) describes the rectilinear motion of a fluid with a constant pressure gradient along the flow direction.

If the fluid is limited by solid walls $b = 0$ and $b = h$, then due to the condition of sticking to solid surfaces, it follows that the flow velocity field in the channel can be represented as

$$U = \frac{1}{2\nu\rho}\frac{dp}{da}b(b - h).$$

If one wall is at rest (let it be the wall $b = 0$) and the other moves in its plane with constant velocity U_0, then Eq. (14.22) will give the following representation for velocity:

$$U = \frac{b}{h}U_0 - \frac{1}{2\nu\rho}\frac{dp}{da}b(h - b).$$

The representation describes the so-called Couette flow. In particular, in the absence of a pressure drop, we obtain a linear velocity distribution

$$U = \frac{b}{h}U_0.$$

A flow with such a velocity distribution is often called a simple Couette flow or a pure shear flow. The Couette flow in a broader sense (with a pressure drop) is the superposition of a simple Couette flow on the flow in the channel. The shape of the velocity distribution curve for the Couette flow is determined by a dimensionless pressure gradient

$$P = -\frac{h^2}{2\nu\rho U_0}\frac{dp}{da}.$$

For $P > 0$, when the pressure drops in the direction of the upper wall motion, the velocity remains positive over the entire channel

Basic Equations and Examples

width. At $P < -1$, a return flow occurs near the stationary wall. This is explained by the fact that the entrainment effect on fluid particles by their neighbors from nearby layers is unable to overcome the pressure drop acting in the direction opposite to the upper wall motion.

14.3.2. *Hagen–Poiseuille flow in a round tube*

Consider an axisymmetric analog of the flow in a channel. Let the fluid particles move along the axis of a circular tube of radius r_0, and their velocity depends on the distance to the axis. Their coordinates change according to the following law:

$$X = a + U(r_\perp)t, \quad y = b, \quad Z = c, \quad r_\perp = \sqrt{b^2 + c^2}. \quad (14.23)$$

Such a representation, as can be easily seen, satisfies the continuity equation (see Eq. (14.6)). The Jacobi matrix for a given flow is written as

$$\hat{R} = \begin{pmatrix} 1 & U'_b & U'_c \\ 0 & 1 & 0 \\ 0 & 0 & 1 \end{pmatrix}; \quad U'_b = U' \frac{b}{r_\perp}; \quad U'_c = U' \frac{c}{r_\perp}; \quad U' = \frac{dU}{dr_\perp},$$

and the flow vorticity, Cauchy invariants, and metric tensor are equal to

$$\Omega_x = 0, \quad \Omega_y = U'_c, \quad \Omega_z = -U'_b,$$
$$S_1 = 0, \quad S_2 = U'_c, \quad S_3 = -U'_b, \quad \text{and} \quad (14.24$$
$$g_{ij} = \begin{pmatrix} 1 & U'_b t & U'_c t \\ U'_b t & 1 + U'^2_b t^2 & U'_b U'_c t^2 \\ U'_c t & U'_b U'_c t^2 & 1 + U'^2_c t^2 \end{pmatrix},$$

respectively. These formulas are the generalization of Eqs. (14 and (14.21). Now, two Cauchy invariants are non-zero, but they coincide with the corresponding vorticity components. Note also the relation $\hat{g}\vec{S} = \vec{S}$ is fulfilled.

256   *Analytical Fluid Dynamics in Lagrangian Variables*

Substitute Eq. (14.24) into Eq. (14.16), from which it follows that the function U satisfies the equation

$$\left(\frac{(r_\perp U')\prime}{r_\perp}\right)' = 0.$$

On solving it, we find that

$$U = \alpha\frac{r_\perp^2}{4} + \beta\ln r_\perp + \gamma,$$

where α, β, and γ are constants. Due to the limited speed, we should assume $\beta = 0$ on the tube axis. The quantity α, as follows from Eqs. (14.6) and (14.23), is equal to

$$\alpha = \frac{1}{\nu\rho}\frac{dp}{da} = U''_{bb} + U''_{cc} = \frac{(r_\perp U')'}{r_\perp} = \text{const},$$

and the constant γ is determined by the condition of sticking on the wall $(U = 0$ at $r = r_0)$. As a result, the flow in a round tube is described by the following formula:

$$U = \frac{1}{4\nu\rho}\frac{dp}{da}(r_\perp^2 - r_0^2).$$

A flat wall suddenly set in motion (Stokes' first problem)

rn to unsteady flows with rectilinear trajectories of fluid
a flat wall that was previously at rest suddenly start to
vn plane with a constant velocity U_0. Let us determine
by such a motion.

hat the wall coincides with the xz plane and moves
of the x-axis. We will seek a solution in the

$$+ f(b,t), \quad Y = b, \quad \text{and} \quad Z = c. \qquad (14.25)$$

20)
still
that

the continuity equation for an arbitrary func-
uation for Cauchy invariants imposes restric-
expressions for $\hat{R}, \vec{S}$, and $\hat{g}$ are written as

follows:

$$\hat{R} = \begin{pmatrix} 1 & f_b & 0 \\ 0 & 1 & 0 \\ 0 & 0 & 1 \end{pmatrix}; \quad \vec{S} = \begin{pmatrix} 0 \\ 0 \\ -f_{tb} \end{pmatrix}; \quad \hat{g} = \begin{pmatrix} 1 & f_b & 0 \\ f_b & 1 + f_b^2 & 0 \\ 0 & 0 & 1 \end{pmatrix}.$$

$$(14.26)$$

Substituting these relations into Eq. (14.18), we find the equation for the function f:

$$f_{ttb} = \nu f_{tbbb}.$$

Let us integrate this equation by the Lagrangian variable:

$$f_{tt} = \nu f_{tbb} + \beta(a, t). \tag{14.27}$$

From Eq. (14.6), it is not difficult to determine that $\beta(a, t) = -\rho^{-1} p_a$. We assume that the pressure in the entire space is constant. Then, $\beta = 0$, and Eq. (14.27) has the form of a heat conduction equation for the horizontal velocity $X_t = f_t = u(b, t)$:

$$u_t = \nu u_{bb}. \tag{14.28}$$

The initial conditions will be

$$\left. \begin{array}{llll} \text{at } t \leq 0: & u = 0 & \text{for all } b, \\ \text{at } t > 0: & u = U_0 & \text{for } b = 0; \\ & u = 0 & \text{for } b = \infty. \end{array} \right\}$$

By introducing a dimensionless variable

$$\eta = \frac{b}{2\sqrt{\nu t}}$$

and assuming $u = U_0 g(\eta)$, we reduce the partial differential Eq. (14.28) to an ordinary differential equation

$$g'' + 2\eta g' = 0$$

with boundary conditions

$$g = 1 \text{ at } b = 0 \quad \text{and} \quad g = 1 \text{ at } b = \infty.$$

258   *Analytical Fluid Dynamics in Lagrangian Variables*

The solution to this equation will be

$$u = U_0 \, erfc \, \eta,$$

where

$$erfc \, \eta = \frac{2}{\sqrt{\pi}} \int_\eta^\infty \exp(-\eta^2)d\eta = 1 - erf\eta = 1$$
$$-\frac{2}{\sqrt{\pi}} \int_0^\eta \exp(-\eta^2)d\eta$$

is an additional probability integral.

The only non-zero Cauchy invariant (see Eq. (14.26)) will be equal to

$$S_3 = \frac{U_0}{\sqrt{\pi \nu t}} \exp\left(-\frac{b^2}{4\nu t}\right).$$

In contrast to the previously discussed examples, the Cauchy invariant now depends on time.

14.3.4.   *Flow created by a vibrating plane (Stokes' second problem)*

Let an unlimited flat wall make rectilinear harmonic oscillations with a frequency of ω in its plane. As in the previous problem, we direct the x-axis along the wall and the y-axis perpendicular to the wall. The pressure in the fluid is assumed to be uniformly distributed, and the fluid motion continues to be described through Eq. (14.25). Since the fluid sticks to it, the condition

$$u(b, t)|_{b=0} = u(0, t) = U_0 \cos \omega t$$

is fulfilled for the horizontal velocity $u(b, t)$ on its surface. This condition acts as a boundary condition for Eq. (14.28). Both are satisfied by the following expression:

$$u(b, t) = U_0 e^{-kb} \cos(\omega t - kb), \quad k = \sqrt{\frac{\omega}{2\nu}}.$$

Basic Equations and Examples

Thus, the fluid particles near the wall perform an oscillatory motion with decreasing amplitude $U_0 \exp(-kb)$ as they move away from the wall; moreover, the oscillation of the fluid layer located at a distance b from the wall lags in phase from the wall vibrations by $b\sqrt{\omega}/2\nu$.

The Cauchy invariant for this type of flow is

$$S_3 = kU_0 e^{-kb}[\cos(\omega t - kb) - \sin(\omega t - kb)].$$

So far, when seeking an exact solution, the equations for Cauchy invariants have been analyzed first. They do not include pressure. To calculate it, it is necessary to refer to the conventional momentum equations in Lagrangian variables. With allowance for this, we will immediately study the following examples within the framework of the system (14.7).

14.3.5. *Flow between two rotating cylinders*

Consider the motion between two coaxial cylinders rotating at different but constant angular velocities. Let the radii of the inner and outer cylinders be R_1 and R_2, respectively, and the angular velocities of their rotation be ω_1 and ω_2. It is convenient to consider such a motion in the polar coordinate system. Let us denote the Euler polar coordinates R and Φ, and the Lagrangian coordinates r and φ, so that

$$X = R\cos\Phi, \quad Y = R\sin\Phi, \quad a = r\cos\varphi, \quad \text{and} \quad b = r\sin\varphi.$$

In the new variables, system (14.7) can be written as follows:

$$\frac{D(X,Y)}{D(a,b)} = \frac{D(X,Y)}{D(R,\Phi)}\frac{D(R,\Phi)}{D(r,\varphi)}\frac{D(r,\varphi)}{D(a,b)} = \frac{R}{r}\frac{(R,\Phi)}{D(r,\varphi)}$$

$$= \frac{R}{r}[R,\Phi] = 1; \tag{14.29}$$

$$r\frac{\partial^2}{\partial t^2}R\cos\Phi = -\frac{1}{\rho}\,[p, R\sin\Phi]$$

$$+ \nu\left\{\left[R\cos\Phi, \frac{1}{r}\left[R\cos\Phi, \frac{\partial}{\partial t}R\cos\Phi\right]\right]\right\} \tag{14.30}$$

$$+\left[R\sin\Phi,\;\frac{1}{r}\left[R\sin\Phi,\;\frac{\partial}{\partial t}R\cos\Phi\right]\right]\right\};$$

$$r\frac{\partial^2}{\partial t^2}R\sin\Phi = -\frac{1}{\rho}\left[R\cos\Phi,p\right]$$

$$+\nu\left\{\left[R\cos\Phi,\;\frac{1}{r}\left[R\cos\Phi,\;\frac{\partial}{\partial t}R\sin\Phi\right]\right]\right.\qquad(14.31)$$

$$+\left.\left[R\sin\Phi,\;\frac{1}{r}\left[R\sin\Phi,\;\frac{\partial}{\partial t}R\sin\Phi\right]\right]\right\}.$$

Here, square brackets already denote the Jacobian in variables r and φ. When deriving these equations from system (14.7), the "rule" of multiplication of Jacobians was used, which is reflected in writing continuity Eq. (14.29). We will be interested in flows with circular streamlines where

$$R = r, \quad \Phi = \varphi + f(r,t), \quad \text{and} \quad p = p(r,t).$$

The continuity equation in this form of solution is satisfied for any function f. Substituting these relations into momentum Eqs. (14.31) and (14.32) will lead to the following conditions for the choice of f:

$$\rho r f_t^2 = \frac{dp}{dr};$$
$$\qquad(14.32)$$
$$r f_{tt} = \nu(r f_{trr} + 3 f_{tr}).$$

Since f_t is the angular velocity of rotation of fluid particles, the boundary conditions for this system have the form

$$f_t|_{r=R_1} = \omega_1 \quad \text{and} \quad f_t|_{r=R_2} = \omega_2. \qquad(14.33)$$

We assume f to be a linear function of time. Then, it follows from Eq. (14.32) that

$$r f_{trr} + 3 f_{tr} = 0,$$

and the expression for the angular velocity of rotation of fluid particles satisfying condition (14.33) has the following form:

$$f_t = \frac{\omega_2 R_2^2 - \omega_1 R_1^2}{R_2^2 - R_1^2} + \frac{(\omega_1 - \omega_2)\,R_1^2 R_2^2}{(R_2^2 - R_1^2)r^2}.$$

Basic Equations and Examples 261

Using Eq. (14.12), we find the Cauchy invariant

$$S_3 = \frac{1}{r}\frac{\partial}{\partial r}(r^2 f_t) = \frac{2(\omega_2 R_2^2 - \omega_1 R_1^2)}{R_2^2 - R_1^2}.$$

It does not depend on time due to the stationary nature of the flow. The same result could be obtained by calculating the vorticity.

14.3.6. *Vortex diffusion*

By direct substitution, one can verify that the expression

$$f_t = \frac{\Gamma_0}{2\pi r^2}\left[1 - \exp\left(-\frac{r^2}{4\nu t}\right)\right] \tag{14.34}$$

is an exact solution of Eq. (14.32). It describes the diffusion of vorticity from a vortex thread with an initial circulation of Γ_0. This solution was obtained by Hamel in 1916 and independently by Oseen in 1927 [6]. The Cauchy invariant for the flow (14.34) is calculated by Eq. (14.12). The Cauchy invariant is equal to

$$S_3 = \frac{\Gamma_0}{4\pi\nu t}\exp\left(-\frac{r^2}{4\nu t}\right).$$

At the initial time $(t = 0)$, the value S_3 has a singularity (the vorticity is equal to infinity). Subsequently, the point vortex becomes more and more blurred. The vorticity on the axis becomes finite, but it still remains maximum at this moment.

14.3.7. *Diffuser flow*

Consider the motion of a viscous fluid between two flat walls inclined to each other at an angle α. Assume that the flow is purely radial and is given in the Lagrangian representation by the following relations:

$$\Phi = \varphi; \quad R^2 = r^2 + 2f(\varphi)t; \quad p(r,\varphi) = \frac{2\nu\rho}{R^2}\left[f(\varphi) - \frac{C}{4}\right] + C_1, \tag{14.35}$$

where C, and C_1 are constants. On substituting them into Eqs. (14.31) and (14.32), we obtain the equation for f:

$$f'' + 4f + \frac{f^2}{\nu} - C = 0. \tag{14.36}$$

262   *Analytical Fluid Dynamics in Lagrangian Variables*

On the diffuser walls, where $\varphi = \pm\alpha/2$, the sticking condition

$$f(\pm\alpha/2) = 0 \tag{14.37}$$

is fulfilled. We should add the condition of constant flow of fluid through the cross section of the diffuser per unit time

$$Q = \rho \int_{-\alpha/2}^{\alpha/2} V_R \cdot R d\varphi = \rho \int_{-\alpha/2}^{\alpha/2} f(\varphi) d\varphi. \tag{14.38}$$

The quantity Q is called the abundance of the source and is considered a given value. If $Q > 0$, we are dealing with a source, i.e., a divergent source in the diffuser. If Q is negative, we are dealing with a sink (converging flow or flow in the confuser).

Equation (14.36) can be easily integrated [7]. Its general solution is written in terms of the Weierstrass elliptic function. A monograph [8] provided a complete analysis of the solution when conditions (14.37) and (14.38) are imposed. Divergent flow in the diffuser cannot occur with a large number Re (its role in this problem is played by $|Q|/\nu\rho$). If a certain Reynolds value is exceeded, the fluid flow out of the diffuser can only occur in such a way that there are inflow areas. A convergent, axially symmetric flow is possible for any Reynolds number.

Substituting Eq. (14.35) into Eq. (14.12), we find that

$$\vec{S} = -\frac{f'}{r^2 + 2ft}\vec{c}^0 = -\frac{f'}{R^2}\vec{c}^0.$$

The invariant S_3 coincides with the vorticity at the point where this fluid particle is located.

References

[1]   Gerber, R. (1949). Sur la reduction a un principe variationnel des equations du movement d'un fluide visqueux incompressible. *Ann. Inst. Fourier*, 1, pp. 157–162.

[2]   Monin, A.S. (1962). On the Lagrangian equations of the hydrodynamics of an incompressible viscous fluid. *J. Appl. Math. Mech.*, 26(2), pp. 458–468.

[3]   Monin, A.S. and Yaglom, A.M. (1971). *Statistical Fluid Mechanics*, Vol. 1 (MIT Press, Cambridge, Mass).

Basic Equations and Examples 263

[4] Abrashkin, A.A. and Yakubovich, E.I. (2007). New form of the equations of motion of a viscous fluid in Lagrangian variables. *J. Appl. Mech. Tech. Phys.*, 48(2), pp. 153–158.

[5] Andreev, V.K. (1992). *Stability of Unsteady Fluid Motion with a Free Boundary* [in Russian], (Nauka, Novosibirsk).

[6] Schlichting, H. (1960). *Boundary Layer Theory* (McGraw-Hill, New York).

[7] Hamel, G. (1916). Spiralformige Bewegungen zaher Flussigkeiten. *Jahresbericht der deutschen Mathematiker Vereinigung*, 25, pp. 34–60.

[8] Kochin, N.E., Kibel, I.A. and Roze, N.V. (1964). *Theoretical Hydromechanics*, Vol. 1 (Interscience Publ., New York).

Chapter 15

Gravity Waves on the Surface of a Viscous Fluid

"Who, the waves, stopped you,
Who has shackled your mighty run ..."

— A.S. Pushkin

All viscous flows considered in the previous chapter were initially obtained in Eulerian coordinates. By their example, we have directly verified that calculations within the Lagrangian approach are more cumbersome. However, the situation regarding the description of surface waves in a viscous fluid is different.

The point is that the boundary layer thickness near the free surface is small compared to the wave amplitude. As a consequence, it is incorrect to set the boundary conditions at $y = 0$ (y is the vertical coordinate). The choice of a reference frame moving with the wave and the use of an orthogonal curvilinear coordinate grid in which the free boundary is a coordinate surface are the most optimal. In the Eulerian description, this type of surface is unknown, so the boundary conditions are formulated on a plane surface $y = 0$ when linear plane waves are studied and on a sinusoidal surface, which corresponds to the profile of a linear wave [1–3], when quadratic effects are considered. The study of the following approximations or nonlinear spatial waves with this approach is extremely cumbersome and almost unrealistic since a significant number of calculations are required.

266   *Analytical Fluid Dynamics in Lagrangian Variables*

However, these difficulties are absent in the Lagrangian description of wave motion. The shape of the free surface is given by the condition $b = 0$ (b is the vertical Lagrangian coordinate), and it depends neither on the order of perturbation theory nor on the dimensionality of the problem. Within the framework of the Lagrangian approach, a number of calculations for two-dimensional surface waves were carried out [4, 5]. In particular, the influence of various factors, such as air motion, Earth's rotation [4], and elastic films on the free surface [5], on the drift flow in a fluid has been elucidated. The technique of applying the Lagrangian method for different types of surface waves is described in the following.

15.1. Standing Waves

15.1.1. *Linear spatial oscillations* [6]

Consider a spatial standing gravity-capillary wave in an infinitely deep, viscous fluid. The system of equations in Lagrangian variables a, b, and c has the form (14.6)

$$[X, Y, Z] = \frac{D(X, Y, Z)}{D(a, b, c)} = 1; \tag{15.1}$$

$$X_{tt} = -\frac{1}{\rho}[Y, Z, p] + \nu \left\{ [Y, Z, [Y, Z, X_t]] + [Z, X, [Z, X, X_t]] \right.$$
$$\left. + [X, Y, [X, Y, X_t]] \right\};$$

$$Y_{tt} = -g - \frac{1}{\rho}[Z, X, p] + \nu \left\{ [Y, Z, [Y, Z, Y_t]] + [Z, X, [Z, X, Y_t]] \right.$$
$$\left. + [X, Y, [X, Y, Y_t]] \right\};$$

$$Z_{tt} = -\frac{1}{\rho}[X, Y, p] + \nu \left\{ [Y, Z, [Y, Z, Z_t]] \right.$$
$$\left. + [Z, X, [Z, X, Z_t]] + [X, Y, [X, Y, Z_t]] \right\}. \tag{15.2}$$

Here, the gravity force in the vertical direction was taken into account.

Systems (15.1)–(15.2) should be supplemented with boundary conditions. In the case of free waves in deep water, this is the requirement for no leakage at the bottom ($Y_t = 0$ at $b = -\infty$) and the

Gravity Waves on the Surface of a Viscous Fluid   267

dynamic condition on the free surface, which can be written as follows:

$$T_{ik}n_k = (-p_0 + \gamma K)n_i; \quad b = 0;$$

$$\vec{n} = \frac{\vec{x}_0(Y_cZ_a - Y_aZ_c) + \vec{y}_0(Z_cX_a - Z_aX_c) + \vec{z}_0(X_cY_a - X_aY_c)}{\sqrt{(Y_aZ_c - Y_cZ_a)^2 + (Z_aX_c - Z_cX_a)^2 + (X_aY_c - X_cY_a)^2}};$$

$$K = \frac{Y_{aa}(1 + Y_c^2) - 2Y_aY_cY_{ac} + Y_{cc}(1 + Y_a^2)}{(1 + Y_a^2 + Y_c^2)^{3/2}}, \tag{15.3}$$

where T_{ik} is the viscous stress tensor, p_0 is the constant external pressure, γ is the surface stress coefficient, $\vec{n}$ is the external normal to the free surface $Y = Y(a, 0, c)$, and K is its average curvature. The expressions tensor Tik are obtained by the representing them in Jacobian form and using the Jacobian "multiplication" rule, taking into accoount the equation (15.1):

$$T_{xx} = -p + 2\nu\rho[X_t, Y, Z]; \quad T_{yy} = -p - 2\nu\rho[Y_t, X, Z];$$

$$T_{zz} = -p - 2\nu\rho[Z_t, Y, X];$$

$$T_{xy} = T_{yx} = \nu\rho([Y_t, Y, Z] - [X_t, X, Z]);$$

$$T_{xz} = T_{zx} = \nu\rho([Z_t, Y, Z] - [X_t, Y, X]); \tag{15.4}$$

$$T_{yz} = T_{zy} = \nu\rho([Y_t, X, Y] - [Z_t, X, Z]).$$

Let us represent all the functions included in the momentum equations as a series with respect to a small parameter of the wave steepness $\varepsilon = kA$, where k is the wave number and A is the wave amplitude:

$$X = a + \varepsilon\xi_1 + O(\varepsilon^2); \quad Y = b + \varepsilon\eta_1 + O(\varepsilon^2); \quad Z = c + \varepsilon\varsigma_1 + O(\varepsilon^2);$$

$$p = p_0 - \rho g b + \varepsilon p_1 + O(\varepsilon^2).$$

Substituting these relations into Eqs. (15.1)–(15.4) yields equations for unknown functions and expressions for boundary conditions.

The system of equations in the linear approximation has the following form:

$$\xi_{1a} + \eta_{1b} + \varsigma_{1c} = 0;$$

$$\xi_{1tt} = -\rho^{-1}p_{1a} - g\eta_{1a} + \nu\Delta\xi_{1t};$$

$$\eta_{1tt} = -\rho^{-1}p_{1b} - g\eta_{1b} + \nu\Delta\eta_{1t}; \qquad (15.5)$$

$$\varsigma_{1tt} = -\rho^{-1}p_{1c} - g\eta_{1c} + \nu\Delta\varsigma_{1t},$$

where Δ is the Laplacian of the Lagrange variables. The boundary conditions on a free surface $(b = 0)$ for it will be written as

$$\eta_{1ta} + \xi_{1tb} = 0; \quad \eta_{1tc} + \varsigma_{1tb} = 0; \quad -p_1 + 2\nu\rho\eta_{1tb} = \gamma(\eta_{1aa} + \eta_{1cc}). \qquad (15.6)$$

We will seek a solution by the method of separation of variables, assuming

$$\xi_1 = \text{Re}A(b)e^{nt}\sin ka \cos mc; \quad \eta_1 = \text{Re}B(b)e^{nt}\cos ka \cos mc;$$

$$\varsigma_1 = \text{Re}C(b)e^{nt}\cos ka \sin mc; \quad p_1 = \text{Re}H(b)e^{nt}\cos ka \cos mc. \qquad (15.7)$$

The functions A, B, C, H, and n are considered complex. Performing simple algebraic calculations, we arrive at the equation for the function A:

$$A^{IV} - \left[2(k^2 + m^2) + \frac{n}{\nu}\right]A^{II} + \left(k^2 + m^2 + \frac{n}{\nu}\right)A = 0.$$

Assuming $A = \exp(lb)$, we obtain a biquadrate equation for l, the solution of which will be the relations

$$l_1^2 = k^2 + m^2 = M^2; \quad l_2^2 = k^2 + m^2 + \frac{n}{\nu} = N^2.$$

The wave disturbances should decrease with increasing depth (at $b \to -\infty$); therefore, A should be chosen in the form

$$A = \alpha e^{Mb} + \beta e^{Nb}; \quad M, \text{Re}N > 0. \qquad (15.8)$$

Gravity Waves on the Surface of a Viscous Fluid   269

The functions B, C, and H are obtained from system (15.5) and have the form

$$B = -\frac{M}{k}\alpha e^{Mb} + \delta e^{Nb}; \quad C = \frac{m}{k}e^{Mb} + Re^{Nb},$$

$$H = \frac{\rho}{k}(gM + n^2)\alpha e^{Mb} - \rho g\delta e^{Nb}; \tag{15.9}$$

α, β, δ, and R are complex constants, which satisfy the condition

$$k\beta + N\delta + mR = 0, \tag{15.10}$$

which is a consequence of continuity Eq. (15.1). To determine the specific form of these constants, we substitute Eqs. (15.8) and (15.9) into boundary condition (15.6). As a result, we obtain three equations for determining the four constants:

$$2M\alpha + N\beta - k\delta = 0;$$

$$2M\frac{m}{k}\alpha + NR - m\delta = 0;$$

$$k^{-1}[n^2 + 2\nu n M^2 + gM + \rho^{-1}\gamma M^3]\alpha - [g + 2\nu n N + \rho^{-1}\gamma M^2]\delta = 0.$$

Taking into account Eq. (15.10), the compatibility condition of this system will be written as

$$(n + 2\nu M^2)^2 + gM + \rho^{-1}\gamma M^3 = 4\nu^2 M^3 N. \tag{15.11}$$

We introduce the notations

$$\omega^2 = gM + \rho^{-1}\gamma M^3; \quad \frac{\nu M^2}{\omega} = \theta; \quad n + 2\nu M^2 = s\omega; \tag{15.12}$$

taking these into account, Eq. (15.11) can be reduced to

$$(s^2 + 1)^2 = 16\theta^3(s - \theta). \tag{15.13}$$

This expression coincides with the equation for plane surface waves in a viscous fluid; only the role of the wave number in a two-dimensional wave is now played by the quantity $M = \sqrt{k^2 + m^2}$, where k is the longitudinal wave number and m is the transverse wave number (spatial modulation period). The value $m = 0$ corresponds to a plane wave. The expression for the parameter ω coincides with the frequency of gravity-capillary waves propagating over the surface of an ideal fluid.

The constants included in the solution for linear waves are determined by the equalities

$$\beta = -\frac{2MN}{M^2 + N^2}\alpha, \quad \delta = \frac{2M^3}{k(M^2 + N^2)}\alpha, \quad \text{and}$$

$$R = -\frac{2mMN}{k(M^2 + N^2)}\alpha. \tag{15.14}$$

Relations (15.7)–(15.9) and (15.12)–(15.14) give a complete solution to the problem of standing spatial waves on the surface of a viscous fluid. The value α included in Eq. (15.14) specifies the wave amplitude B_0, which is equal to

$$B_0 = |B|\big|_{b=0} = \text{Re}\frac{nM\alpha}{\nu k(M^2 + N^2)}.$$

The quantity $\varepsilon = MB_0$ serves as the wave steepness parameter.

Specific values of the constants in Eq. (15.14) are found by the well-known solution of the dispersion relation (15.13). It makes it possible to find the number s, and hence the value of n, included in the solution to Eq. (15.7), from given values of the fluid viscosity, frequency ω, and wave number M, which determine the value of the parameter θ.

15.1.2. *Long-wave approximation*

We will consider wave perturbations for which

$$M^{-1} \gg \sqrt{2\nu/\omega} = \Delta^{-1}. \tag{15.15}$$

The quantity Δ^{-1} characterizes the thickness of the boundary layer (which is very thin), and the inequality (15.15) itself reflects the fact that the wavelength significantly exceeds this characteristic scale. It follows from Eq. (15.15) that $\theta \ll 1$. If this condition is fulfilled, the right-hand side of Eq. (15.13) can be neglected, and its solution will be the values $s = \pm i$ or, in dimensional variables,

$$n = -2\nu M^2 \pm i\omega. \tag{15.16}$$

Hence, we conclude that the magnitude of the decay rate $\operatorname{Re} n = -2\nu M^2$, and the oscillation frequency is equal to the frequency ω of gravity-capillary waves in an ideal fluid. The sign before the frequency determines the oscillation phase and it can be anything; we choose the plus sign for definiteness.

Taking into account inequality (15.15), expression (15.14) will be rewritten as follows:

$$\beta \approx \frac{2i\nu MN}{\omega}\alpha; \quad \delta \approx -\frac{2i\nu M^3}{\omega k}\alpha;$$
$$R \approx \frac{2i\nu m MN}{\omega k}\alpha; \quad N \approx \Delta(1+i). \tag{15.17}$$

It follows from the last equality that the second terms in Eqs. (15.8) and (15.9) for the functions A, B, C, and H are essential only in a surface boundary layer of thickness Δ^{-1}. The quantity $\operatorname{Re}\alpha$ is equal to $kM^{-1}B_0$. Equations (15.7)–(15.9), with allowance for Eqs. (15.15)–(15.17), describe a spatial weakly damped spatial standing wave.

The formulas for the vorticity vector components have the form

$$\Omega_X = [Z_t, Z, X] + [Y_t, Y, X] = \varepsilon(\varsigma_{1tb} - \eta_{1tc}) + O(\varepsilon^2);$$
$$\Omega_Y = [Z_t, Z, Y] + [X_t, X, Y] = \varepsilon(\xi_{1tc} - \varsigma_{1ta}) + O(\varepsilon^2);$$
$$\Omega_Z = [Y_t, Y, Z] + [X_t, X, Z] = \varepsilon(\eta_{1ta} - \xi_{1tb}) + +O(\varepsilon^2).$$

Substituting representations for the trajectories of fluid particles 15.7 into these formulas, we obtain the following representations for the vorticity components of the first order of perturbation theory:

$$\Omega_{X1} = -\operatorname{Re}\frac{2mMn^2\alpha}{\nu k(M^2 + N^2)}\exp(Nb + nt); \quad \Omega_{Y1} = 0;$$

$$\Omega_{Z1} = \operatorname{Re}\frac{2Mn^2\alpha}{\nu(M^2 + N^2)}\exp(Nb + nt)\sin ka\cos mc.$$

In the long-wave approximation, these expressions will take the form

$$\Omega_{X1} = \frac{2mM\omega}{k}\operatorname{Re} i\alpha\,\exp[\Delta(1+i)b + i\omega t]\cos ka\sin mc;$$
$$\Omega_{Z1} = 2M\omega\operatorname{Re} i\alpha\,\exp[\Delta(1+i)b + i\omega t]\sin ka\cos mc.$$

272   *Analytical Fluid Dynamics in Lagrangian Variables*

The case of $m = 0$ corresponds to a plane wave [7]. Assuming $\Delta = 0$ and $\nu = 0$, we obtain a solution for the spatial potential standing wave [8].

15.1.3. *Nonlinear two-dimensional waves*

In a paper [9], the authors considered the problem of two-dimensional standing waves in a viscous fluid of infinite depth. But the solution had already been obtained in the quadratic approximation. The calculations are cumbersome. In the second order of perturbation theory with respect to a small parameter of the wave steepness, ten complex functions have to be calculated. However, the very fact of using Lagrangian coordinates is important.

As already noted, the boundary layer thickness near the free surface can be much smaller than the wave amplitude. In the Eulerian description, the displacement of the boundary condition to the level of the horizontal surface (unperturbed fluid) is possible for low-viscosity fluids in the quadratic and higher approximations. Thus, the features of the drift flow initiated by periodic waves in a viscous fluid are studied [10] in the quadratic approximation. The parameters of a "viscous" Stokes wave are calculated [11] and an example of the calculation of high-frequency Faraday ripples is given [12] (both calculations are performed in the cubic approximation). In the case of arbitrary viscosity in the quadratic approximation, one passes to curvilinear coordinates parameterizing the profile of a quasi-stationary linear wave [1–3]. While for traveling waves this approach is ineffective starting from the cubic approximation, for standing waves, due to the non-stationarity of the free surface, it is inapplicable in the second order of perturbation theory. In Lagrangian coordinates, such difficulties do not arise.

Let us give a brief overview of the wave solution. The coordinates of the trajectory are determined with accuracy up to the cube of a small parameter:

$$X = a + \varepsilon\xi_1 + \varepsilon^2\xi_2 + O(\varepsilon^3); \quad Y = b + \varepsilon\eta_1 + \varepsilon^2\eta_2 + O(\varepsilon^3).$$

In the long-wave limit, the linear solution is written as

$$\eta_1 = \alpha_0 e^{-2\nu k^2 + kb}\cos ka \sin \omega t;$$

$$\xi_1 = -\alpha_0 e^{-2\nu k^2 + kb}\sin ka \sin \omega t. \tag{15.18}$$

Gravity Waves on the Surface of a Viscous Fluid 273

This is a consequence of the calculations in the previous section for $m = 0$. Here, we chose $\alpha = i\alpha_0$, where α_0 is a real number, so that at zero viscosity, the wave perturbations had a form similar to the solution for potential waves [8]. The quantity $\varepsilon\alpha_0$ is equal to the wave amplitude.

The coordinates of the trajectory of a fluid particle in a linear wave satisfy the equation

$$Y - b = -(X - a)\cotan ka.$$

As in a potential wave, fluid particles move relative to the equilibrium position $X_0 = a$ and $Y_0 = b$ in a straight line inclined at an angle $-\cotan ka$, but the amplitude of their oscillations decreases exponentially. Particles with Lagrangian coordinates $ka = (\pi/2) \pm \pi n$ (n is an integer) correspond to nodes and move horizontally. In the antinodes (for them, $ka = \pi/2$), the particles move vertically.

The flow in the near-surface layer is vortical, and the vorticity varies according to the law

$$\Omega_{z1} = -2k\omega\alpha_0 e^{-2\nu k^2 t + \Delta b}\sin ka\cos(\Delta b + \omega t).$$

In this case, standing waves serve as one of the rare examples of an analytical representation of a flow with a non-stationary vorticity distribution.

The solutions of the quadratic approximation can be written as follows [9]:

$$\xi_2 = \left[\frac{\Delta e^{\Delta b}}{k\sqrt{2}}\sin\left(\Delta b - \frac{\pi}{4}\right) - \frac{3e^{2\Delta b}}{4}\right.$$

$$\left. - \frac{ke^{\sqrt{2}\Delta b}}{2\Delta}\sin\left(\sqrt{2}\Delta b + 2\omega t + \frac{\pi}{4}\right)\right]k\alpha_0^2 e^{-4\nu k^2 t}\sin 2ka;$$

$$\eta_2 = \left\{\left[(1 - \cos 2\omega t)\,e^{2kb} + \frac{ke^{\Delta b}}{\Delta}\left[\frac{\sqrt{2}}{2}\sin\left(\Delta b + \omega t + \frac{\pi}{4}\right)\right.\right.\right.$$

$$\left.\left.\left. - \sqrt{2}\sin\left(\Delta b + \frac{\pi}{4}\right)\right]\right]\right.$$

$$\left. + \left[4e^{\Delta b}\cos\Delta b + \frac{3ke^{2\Delta b}}{\Delta}\right]\cos 2ka\right\}\frac{k\alpha_0^2}{4}e^{-4\nu k^2 t}. \qquad (15.19)$$

274   *Analytical Fluid Dynamics in Lagrangian Variables*

With viscosity tending to zero and Δ, tending to infinity, these expressions turn into formulas determining standing potential waves [8]. In the depths ($\Delta|b| \ll 1$ and $\nu \neq 0$), Eq. (15.19) can be written as follows:

$$\xi_2^{int} = 0; \quad \eta_2^{int} = \frac{k\alpha_0^2}{4}(1 - \cos 2\omega t)e^{-4\nu k^2 t + 2kb}\cos 2ka.$$

Vertical oscillations inhomogeneous in space are superimposed on the rectilinear motion of fluid particles in a linear wave. With allowance for Eq. (15.18), the coordinates of the trajectories of fluid particles obey the "rule"

$$Y - b = -(X - a)\cotan ka + \frac{k(X - a)^2}{2}(\cotan^2 ka - 1),$$

i.e., the particles now move along the sections of the parabolas, and in the antinodes, they oscillate vertically.

When writing Eq. (15.19), we limited ourselves to terms of order k/Δ. But within the framework of the long-wave approximation, they should also be neglected. As a result, in the boundary layer, the following representation of the flow is valid:

$$\xi_2^{bound} = k\alpha_0^2\left[\frac{\Delta e^{\Delta b}}{k\sqrt{2}}\sin\left(\Delta b - \frac{\pi}{4}\right) - \frac{3e^{2\Delta b}}{4}\right]e^{-4\nu k^2 t}\sin 2ka;$$

$$\eta_2^{bound} = \frac{k\alpha_0^2}{4}[(1 - 2\cos 2\omega t)e^{2kb} + 4e^{\Delta b}\cos \Delta b \cos 2ka]e^{-4\nu k^2 t}.$$

The boundary layer effect is due to the appearance of additional terms that depend on Lagrangian coordinates but do not depend on the oscillation frequency. Qualitatively, their effect can be represented as a nonlinear disturbance of the flow in depth since fluid particles oscillate along more complex curves than sections of parabolas.

For terms of order k/Δ, the dependence on frequency is already present. It follows that for shorter waves, the motion of fluid particles near the free surface becomes more complicated. Determining the exact analytical representations in the boundary layer is quite a difficult problem. Therefore, expression (15.19) can also be used as an approximate description for the finite value of the parameter

k/Δ. Obviously, it makes sense to talk more about the qualitative features of the flow.

The vorticity of the waves in the quadratic approximation is given by the expression

$$\Omega_2 = \frac{D(\xi_{1t}, \xi_1)}{D(a,b)} + \frac{D(\eta_{1t}, \eta_1)}{D(a,b)} + \xi_{2tb} - \eta_{2ta}.$$

After substituting the solutions of the first approximation and relation (15.19) into it, we obtain the following representation:

$$\Omega_2 = \omega k^2 \alpha_0^2 \left[(2\cos \Delta b - \sin \Delta b)e^{\Delta b} + 2e^{\sqrt{2}\Delta b}\cos(\sqrt{2}\Delta b + 2\omega t) \right]$$

$$\times e^{-4\nu k^2 t} + o\left(\frac{k}{\Delta}\right).$$

Vorticity is the sum of two fields, namely, quasi-stationary ($\nu k^2 \ll 1$) and oscillating with double frequency.

15.2. Traveling Spatial Waves

Stokes was the first to theoretically show that two-dimensional waves of small amplitude carry out mass transport on average in the wave direction — the Stokes drift (Section 4.1.3). He performed calculations within the framework of an ideal fluid. Having performed similar calculations for a viscous fluid (in Eulerian variables), Longuet-Higgins showed that, if we neglect the weak wave damping, the drift velocity is, first, independent of the fluid viscosity and, second, exactly twice as large as in an ideal fluid [2, 3]. In the following, the mass transport problem is solved for spatial waves [13].

15.2.1. *Navier–Stokes equations in modified Lagrangian coordinates*

Consider the propagation of a spatial wave in an infinitely deep, viscous fluid. We will describe the motion in modified Lagrangian coordinates $q = a + \sigma(b)t, b, c$ (see Section 4.1.1). The system of

276 *Analytical Fluid Dynamics in Lagrangian Variables*

momentum equations of an incompressible viscous fluid in these variables has the following form:

$$[X, Y, Z] = 1;$$

$$\sigma^2 X_{qq} + 2\sigma X_{qt} + X_{tt}$$

$$= -\frac{1}{\rho}[Y, Z, p] + \nu\{[Y, Z, [Y, Z, X_t + \sigma X_q]]$$

$$+ [Z, X, [Z, X, X_t + \sigma X_q]] + [X, Y, [X, Y, X_t + \sigma X_q]]\};$$

$$\sigma^2 Y_{qq} + 2\sigma Y_{qt} + Y_{tt}$$

$$= -g - \frac{1}{\rho}[Z, X, p] + \nu\{[Y, Z, [Y, Z, Y_t + \sigma Y_q]]$$

$$+ [Z, X, [Z, X, Y_t + \sigma Y_q]] + [X, Y, [X, Y, Y_t + \sigma Y_q]]\}$$

$$\sigma^2 Z_{qq} + 2\sigma Z_{qt} + X_{tt}$$

$$= -\frac{1}{\rho}[X, Y, p] + \nu\{[Y, Z, [Y, Z, Z_t + \sigma Z_q]]$$

$$+ [Z, X, [Z, X, Z_t + \sigma Z_q]] + [X, Y, [X, Y, Z_t + \sigma Z_q]]\}.$$

$$(15.20)$$

Here, square brackets denote the Jacobian taking operation. These equations are derived from Eqs. (15.1) and (15.2). We assume the function $\sigma(b)$ to be constant: $\sigma(b) = \sigma = \text{const}$.

Along with the condition of damping of vertical oscillations at the bottom ($Y_t \to 0$ at $b \to -\infty$), it is necessary to require that the relations

$$T_{ik}n_k = (-p_0 + \gamma K)n_i, \quad b = 0,$$

$$\vec{n} = \frac{\vec{x}_0(Y_c Z_q - Y_q Z_c) + \vec{y}_0(Z_c X_q - Z_q X_c) + \vec{z}_0(X_c Y_q - X_q Y_c)}{\sqrt{(Y_q Z_c - Y_c Z_q)^2 + (Z_q X_c - Z_c X_q)^2 + (X_q Y_c - X_c Y_q)^2}},$$

$$(15.21)$$

and

$$K = \frac{Y_{qq}(1 + Y_c^2) - 2Y_q Y_c Y_{qc} + Y_{cc}(1 + Y_q^2)}{(1 + Y_q^2 + Y_c^2)^{3/2}}$$

be fulfilled on the free surface (see Eqs. (15.3) and (15.4), where T_{ik} is the viscous stress tensor, p_0 is a constant external pressure, γ is the surface tension coefficient, $\overrightarrow{n}$ is the external normal to the free surface $Y = Y(q, 0, c)$, and K is its average curvature. Expressions for the T_{ik} tensor components are written as

$$T_{xx} = -p + 2\nu\rho[X_t + \sigma X_q, Y, Z], \quad T_{yy} = -p - 2\nu\rho[Y_t + \sigma Y_q, X, Z];$$

$$T_{zz} = -p - 2\nu\rho[Z_t + \sigma Z_q, Y, X];$$

$$T_{xy} = T_{yx} = \nu\rho([Y_t + \sigma Y_q, Y, Z] - [X_t + \sigma X_q, X, Z]); \tag{15.22}$$

$$T_{xz} = T_{zx} = \nu\rho([Z_t + \sigma Z_q, Y, Z] - [X_t + \sigma X_q, Y, X]);$$

$$T_{yz} = T_{zy} = \nu\rho([Y_t + \sigma Y_q, X, Y] - [Z_t + \sigma Z_q, X, Z]).$$

The construction of the traveling wave perturbation theory is similar to the case of a standing wave (Section 15.1.1). Linear traveling waves have the same dispersion relation as standing waves. As a consequence, the vertical structure of linear disturbances is similar for both types of waves. According to the known solution for linear standing waves, it is easy to write expressions for the coordinates of the traveling wave trajectories [13].

15.2.2. *Averaged drift*

To calculate the averaged horizontal drift of fluid particles, it is necessary, first of all, to write out the equation for the ξ_2-quadratic perturbation of the horizontal coordinate of a fluid particle. Perturbations of the second order are included in this equation linearly, and various perturbation derivatives $\xi_1, \eta_1, \varsigma_1$, and p_1 form nonlinear combinations. Outside the boundary layer, they are represented as

278   *Analytical Fluid Dynamics in Lagrangian Variables*

follows:

$$\xi_1 = -kM^{-1}B_* \exp(Mb + nt) \cos mc \sin kq;$$

$$\eta_1 = B_* \exp(Mb + nt) \cos mc \cos kq;$$

$$\varsigma_1 = -mM^{-1}B_* \exp(Mb + nt) \sin mc \cos kq;$$

$$p_1 = \rho(\omega^2 - gM)B_* \exp(Mb + nt) \cos mc \cos kq.$$

(15.23)

Here, B_* is a constant that determines the initial wave amplitude, and all other quantities have the same value as in the problem of a standing spatial wave (Section 15.1.1).

When averaging the equation for ξ_2 at the wavelength $2\pi/k$, the terms ξ_{2q}, η_{2q}, and p_{2q}, which are included linearly and are periodic over q, will not contribute to the equation. Taking into account the averaging of nonlinear terms containing derivatives of function (15.23), we come to the following equation for the function $\langle \xi_{2t} \rangle = U$ (angle brackets are an averaging sign):

$$U_t - \nu(U_{bb} + U_{cc}) = -2\nu\omega k M^2 B_*^2 \left[1 + \left(\frac{k}{M}\right)^2 + \left(\frac{k}{M}\right)^4 \cos 2mc \right]$$

$$\exp(2Mb + nt).$$

(15.24)

If the observation time is much less than the wave damping time, the drift velocity can be considered time-independent and the term U_t on the left-hand side of Eq. (15.24) can be neglected. Then, its solution is written in the form

$$U = \frac{\omega k B_*^2}{2M^2}(m^2 + 4k^2 \cos^2 mc)\exp(2Mb + nt).$$

(15.25)

This solution determines the averaged drift of fluid particles in the direction of propagation of a spatial surface wave in a viscous fluid. The velocity U does not depend on the viscosity.

For a potential spatial wave, the drift velocity is [8]

$$U_{pot} = \frac{\omega k B_*^2}{2M^2}(m^2 + 2k^2 \cos^2 mc)\exp 2Mb.$$

The velocity differs from our result (for $nt \ll 1$) only by the multiplier before the square of the cosine.

Gravity Waves on the Surface of a Viscous Fluid 279

For a plane wave, it follows from Eq. (15.25) that

$$U_{pl} = 2\omega k B_*^2 \exp(2Mb + nt). \tag{15.26}$$

This result was first obtained by Longuet-Higgins [2] using the Eulerian approach. The drift velocity in a plane wave is exactly twice the Stokes drift.

Let us compare mass transport in two-dimensional and spatial waves. To do this, we now average Eq. (15.25) over the modulation period. Denoting the averaging operation over the transverse wave scale $2\pi/m$ by two angle brackets, we find

$$\langle\langle U \rangle\rangle = \frac{1}{2}\omega k \frac{m^2 + 2k^2}{m^2 + k^2} B_*^2 \exp(2Mb + nt). \tag{15.27}$$

Comparing Eqs. (15.26) and (15.27), it is easy to show that the average mass transport in a spatial wave is always smaller.

15.3. A Train of Two-Dimensional Gravity Surface Waves for Large Reynolds Numbers

The evolution of a train of two-dimensional potential gravity ($\gamma = 0$) surface waves is described by the nonlinear Schrödinger equation in the cubic order of perturbation theory [14] and the Dysthe equation in the fourth order of asymptotic expansions [15]. In order to account for the viscosity effect, it is customary to add, purely phenomenologically (from general considerations), a linear amplitude term to these evolution equations [16,17]. At the same time, the conditions for correctly introducing an additional term into such equations are not discussed. The magnitude of the multiplier acts as an arbitrary parameter before the amplitude. In this connection, the question arises about the consistent derivation of dissipative evolution equations from the Navier–Stokes equations and the study of the conditions of their applicability.

Consider, following a study [18], the dynamics of a weakly nonlinear train of two-dimensional gravity waves in a viscous fluid. Let us introduce complex variables $W = X + iY$ and $\overline{W} = X - iY$ (the bar is the symbol of complex conjugation); then, the system of Eq. (15.20) for our case can be written as follows:

$$[\overline{W}, W] = \frac{D(\overline{W}, W)}{D(q, b)} = 2i;$$

$$W_{tt} + 2\sigma W_{t\sigma} + \sigma^2 W_{qq} = -ig + \frac{i[p, W]}{\rho}$$
$$+ \frac{\nu}{2}\{[W, [\overline{W}, W_t + \sigma W_q]] \qquad (15.28)$$
$$+ [\overline{W}, [W, W_t + \sigma W_q]]\}.$$

Here, square brackets denote the operation of taking the Jacobian over q and b. We non-dimensionalize this function by introducing new unknown functions and new variables,

$$W = LW_*, \quad p = \rho\sigma^2 p_*, \quad q = Lq_*, \quad \text{and} \quad t = (L/\sigma)t_*, \quad (15.29)$$

where L is some length scale (the subscripts will be omitted in what follows). In dimensionless notations, the system (15.29) will take the form

$$[\overline{W}, W] = 2i;$$
$$W_{tt} + 2W_{t\sigma} + W_{qq} = -i\frac{gL}{\sigma^2} + i[p, W] + \frac{1}{2R_*}\{[W, [\overline{W}, W_t + W_q]]$$
$$+ [\overline{W}, [W, W_t + W_q]]\}. \qquad (15.30)$$

There are two dimensionless parameters in the momentum equation, namely, the Reynolds number $R_* = \sigma L/\nu$ and the quantity gL/σ^2.

The boundary conditions for this system will take the following form (see Eq. (15.21)):

$$T_{ik}n_k = -p_0 n_i, \quad \vec{n}\{n_x, n_y\} = \vec{n}\left\{-\frac{\text{Im}W_q}{|W_q|}, \frac{\text{Re}W_q}{|W_q|}\right\}, b = 0;$$

$$T_{xx} = -p - \frac{i}{R_*}\text{Im}([W_t + W_q, W - \overline{W}]); \qquad (15.31)$$

$$T_{yy} = -p + \frac{i}{R_*}\text{Im}([W_t + W_q, W + \overline{W}]);$$

$$T_{xy} = -\frac{1}{R_*}\text{Re}([W_t + W_q, W]).$$

Gravity Waves on the Surface of a Viscous Fluid 281

Based on Eqs. (15.30) and (15.31), we studied the dynamics of a wave train with slowly varying parameters (amplitude, frequency, and wave number) [18]. The derivative expansion method (method of multiple scales [19]) was employed. It was assumed that the function W depends on the spatial variables $q_0 = q, q_1 = \varepsilon q, q_2 = \varepsilon^2 q$, and b and the time variables $t_1 = \varepsilon t$ and $t_2 = \varepsilon^2 t$ (ε is a small parameter of the wave steepness) and can be represented by the expansion

$$W(q_0, q_1, q_2, b, t_1, t_2) = q_0 + ib + \varepsilon w_1 + \varepsilon^2 w_2$$
$$+ \varepsilon^3 w_3 + O(\varepsilon^4); \tag{15.32}$$
$$w_j = w_j(q_0, q_1, q_2, b, t_1, t_2), j = 1, 2, 3;$$

the pressure in the fluid is determined by the relation

$$p = p_0 - b + \varepsilon p_1 + \varepsilon^2 p_2 + \varepsilon^3 p_3 + O(\varepsilon^4),$$
$$p_j = p_j(q_0, q_1, q_2, b, t_1, t_2). \tag{15.33}$$

Substituting the first two terms (15.33), which define the hydrostatic pressure, into the second equation of system (15.30), we can find the value of the second dimensionless parameter: $gL/\sigma^2 = 1$. If we assume $L = k^{-1}$, and $\sigma = w/k$ is the phase velocity of the wave, this relation will coincide with the dispersion relation for waves in deep water. Note that in the absence of perturbations ($w_0 = q_0 + ib$), the velocity at the bottom, as in the entire fluid, is equal to σ.

In general, it is rather difficult to describe the evolution of a wave train in a viscous fluid. However, let us make a simplifying assumption. Suppose that the Reynolds number is sufficiently large, such that $R_*^{-1} = \kappa \varepsilon^2$, where κ is some numerical coefficient of the order of unity. This relation is equivalent to the fact that the carrier wavelength is $\lambda = (\nu/\kappa\varepsilon^2 \sqrt{g})^{2/3}$. The chosen constraint on the Reynolds number means that viscosity effects will only appear in the cubic approximation.

In each of the approximations, in addition to taking into account the wave component of motion, it is also necessary to determine the average flows. To determine them, it is convenient to use an equation that is obtained by eliminating pressure from the momentum equations. In the variables introduced earlier, this equation can be rewritten in the form

$$\mathrm{Re}\left\{\frac{\partial}{\partial t}[W_t + W_q, \overline{W}] + \frac{\partial}{\partial q}[W_t + W_q, \overline{W}]\right\}$$
$$- \frac{1}{R_*}[W, [\overline{W}, [W_t + W_q, \overline{W}]]] = 0.$$

The vorticity in the complex representation is equal to

$$\Omega = \mathrm{Re}[W_t + W_q, \overline{W}].$$

In the solution process, the vorticity should also be represented as a series of ε.

As a result of solving the formulated problem, it is possible to show that the amplitude A of the envelope of the wave packet satisfies the equation

$$iA_t - \frac{\omega}{8k^2}A_{\xi\xi} - \frac{1}{2}\omega k^2 |A|^2 A + i\frac{\kappa\omega}{R_*}A = 0, \quad \xi = q - \frac{\sigma}{2}t.$$

The most important feature of this equation is that it does not include the drift flow. This is a consequence of the fact that a case of sufficiently large Reynolds numbers is being considered. If the inverse Reynolds number is of the order of the wave steepness or the order of unity, then the evolution equation will already include the mean flow (as the Dysthe equation).

Now, let us consider one drawback of the solution. On the free surface of a viscous fluid, the conditions of continuity of vertical and horizontal components of the momentum flux must be fulfilled. From the first condition, we obtain the nonlinear Schrödinger equation with a dissipative linear term. The second condition cannot be fulfilled. This is due to the fact that near the free surface, it is fundamentally impossible to neglect viscosity. The non-vanishing of the horizontal component of the momentum flux means that the mean flow determined when solving the problem will actually differ from that given by our solution.

To conclude, we make one more remark. To include the dissipative linear term in the evolution equation of order n: the inverse Reynolds number should be of the order of the power $n - 1$ steepness.

References

[1] Philips, O.M. (1977). *The Dynamics of the Upper Ocean* (Cambridge University Press, Cambridge).

[2] Longuet-Higgins, M.S. (1953). Mass transport in water waves. *Phil. Trans. R. Soc. Lond. A*, 245, pp. 535–581.

[3] Longuet-Higgins, M.S. (1960). Mass transport in the boundary layer at a free oscillating surface. *J. Fluid Mech.*, 8, pp. 293–306.

[4] Weber, J.E. and Førland, E. (1990). Effect of the air on the drift velocity of water waves. *J. Fluid Mech.*, 218, 619–640.

[5] Weber, J.E. (1997). Mass transport induced by surface waves in viscous rotating fluid. *Free Surface Flows with Viscosity* (Computational Mechanics: Southampton, Boston).

[6] Abrashkin, A.A. and Bodunova, Yu. P. (2011). Spatial standing waves on the surface of viscous fluid, in: *Proc. of Nizhegorodskii Gosudarstvennyui Tekhnicheskii Universitet*, 87(2), pp. 49–54 [in Russian].

[7] Abrashkin, A.A. and Bodunova, Yu.P. (2010). Gravity waves on the surface of viscous fluid: Lagrange approach. *Phys. Wave Phenom.*, 18(4), pp. 251–255.

[8] Sretensky, L.N. (1977). *The Wave Motion Theory* [in Russian] (Nauka, Moscow).

[9] Abrashkin, A.A. and Bodunova, Yu.P. (2012). Nonlinear standing waves on the surface of viscous fluid. *Fluid Dyn.*, 47, pp. 725–734.

[10] Belonozhko, D.F. and Kozin, A.V. (2011). Features of constructing a drift flow initiated by periodic waves propagating on the surface of viscous fluid. *Fluid Dyn.*, 46, 270–277.

[11] Barinov, V.A. and Basinsky, K.Yu. (2011). Nonlinear Stokes waves on the surface of weakly viscous fluid. *Vestn. Udmurt. Univers.*, 2, pp. 112–122.

[12] Lubimov, D.V., Lubimova, T.P. and Cherepanov, A.A. (2003). *Dynamics of interface surfaces in vibration fields* [in Russian] (Fizmatgiz, Moscow).

[13] Abrashkin, A.A. (2008). Spatial waves on the surface of viscous fluid. *Fluid Dyn.*, 43, pp. 915–922.

[14] Zakharov, V.E. (1968). Stability of periodic waves of finite amplitude on the surface of deep fluid. *J. Appl. Mech. Tech. Phys.*, 9, pp. 190–194.

[15] Dysthe, K.B. (1979). Note on a modification to the nonlinear Shrödinger equation for application to deep water waves. *Proc. Roy. Soc. London A*, 369(1736), pp. 105–114.

284 *Analytical Fluid Dynamics in Lagrangian Variables*

[16] Segur, H., Henderson, D., Carter, J., Hammack, J., Li, C-M., Pheiff, D. and Socha, K. (2005). Stabilizing the Benjamin-Feir instability. *J. Fluid Mech.*, 539, pp. 229–271.

[17] Canney, N.E. and Carter, J.D. (2007). Stability of plane waves on deep water with dissipation. *J. Math. and Comput. Simulation*, 72(2–3), pp. 159–167.

[18] Abrashkin, A.A. and Bodunova, Yu.P. (2013). A packet of gravity surface waves with large Reynolds numbers in viscous fluid. *Fluid Dyn.*, 48, pp. 223–231.

[19] Nayfeh, A. (1973). *Perturbation Methods* (Wiley, New York).

Part VI

The Earth's Rotation Effect

Chapter 16

Exact Solutions for Waves

> "Addressed to a friend,
> The song goes in a circle,
> Because the Earth is round."

> — M. Tanich and I. Shaferan

Let us choose a reference frame on the rotating Earth, as shown in Fig. 16.1. Its origin is at latitude Φ, the X-axis points east, the Y-axis points north, and the Z-axis points vertically upward. In this reference frame, the angular velocity vector of the Earth's rotation $\vec{\Omega}$ lies in the YOZ plane. In a rotating system, in addition to gravity, each particle is affected by the Coriolis force and the centrifugal force, and the momentum equation will be written in the form [1]

$$\vec{R_{tt}} + 2\vec{\Omega} \times \vec{R_t} = -\frac{1}{\rho}\nabla p + \nabla\Phi - \vec{\Omega} \times (\vec{\Omega} \times \vec{R}), \qquad (16.1)$$

where $\Phi = -gZ$ is the gravity potential: Coriolis acceleration is included on the left-hand side of Eq. (16.1) and centripetal acceleration with the minus sign is included on the right-hand side. The centrifugal force has a gradient character, so Eq. (16.1) can be rewritten as follows:

$$\vec{R_{tt}} + 2\vec{\Omega} \times \vec{R_t} = -\frac{1}{\rho}\nabla H; \qquad (16.2)$$

$$H = \frac{p}{\rho} - \Phi + \Phi_c; \quad \Phi_c = -\frac{1}{2}(\vec{\Omega} \times \vec{R})^2,$$

where Φ_c is the potential of centrifugal forces.

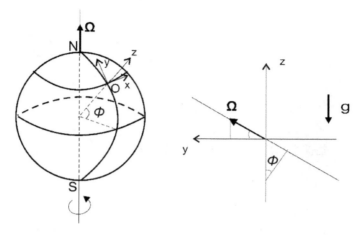

Fig. 16.1. The coordinate system on the Earth's surface.

Multiplying scalar Eq. (16.1) by $\overrightarrow{R_{a_\iota}}$, we obtain the momentum equations in Lagrangian coordinates:

$$\overrightarrow{R_{tt}}\overrightarrow{R_{a_\iota}} + 2(\overrightarrow{\Omega}, \overrightarrow{R_t}, \overrightarrow{R_{a_\iota}}) = -H_{a_i}, i = 1, 2, 3; \{a_i\} = \{a, b, c\}. \quad (16.3)$$

Together with continuity Eq. (1.11), the three equations in Eq. (16.3) form a system of equations of ideal incompressible fluid equations in Lagrangian variables in a rotating reference frame.

In the following subsections, various wave motions will be studied when

(a) the angular velocity of the Earth's rotation in the whole flow region can be considered constant, i.e., the Coriolis parameters $f = 2\Omega_Z = 2\Omega \cdot \sin\Phi$ and $\tilde{f} = 2\Omega_Y = 2\Omega \cdot \cos\Phi$ are considered constant (f-plane approximation);
(b) equatorial flows in the low-latitude band $\Phi \sim Y/R_*$ are studied, where R_* is the Earth's radius, i.e., the Coriolis parameters are $f = \beta Y, \beta = 2\Omega/R$, and $\tilde{f} = 2\Omega$ (β-plane approximation).

The representations for the vector $\overrightarrow{\Omega}$ will be different in each case, but they can be considered within the framework of a general approach [2, 3]. Let us exclude the gradient term from Eq. (16.3) by taking the cross derivatives, and after intermediate calculations, we obtain

$$\frac{\partial}{\partial t}[\overrightarrow{R_{ta_j}}\,\overrightarrow{R_{a_i}} - \overrightarrow{R_{ta_i}}\,\overrightarrow{R_{a_j}} + 2(\overrightarrow{\Omega},\overrightarrow{R_{a_j}},\overrightarrow{R_{a_i}})] = 0, \quad i \neq j.$$

This equation is equivalent to the conditions for preserving three invariants, S_1, S_2, and S_3,

$$\overrightarrow{R_{tb}}\,\overrightarrow{R_c} - \overrightarrow{R_{tc}}\,\overrightarrow{R_b} + 2(\overrightarrow{\Omega},\overrightarrow{R_b},\overrightarrow{R_c}) = S_1(a,b,c); \qquad (16.4)$$

$$\overrightarrow{R_{tc}}\,\overrightarrow{R_a} - \overrightarrow{R_{ta}}\,\overrightarrow{R_c} + 2(\overrightarrow{\Omega},\overrightarrow{R_c},\overrightarrow{R_a}) = S_2(a,b,c); \qquad (16.5)$$

$$\overrightarrow{R_{ta}}\,\overrightarrow{R_b} - \overrightarrow{R_{tb}}\,\overrightarrow{R_a} + 2(\overrightarrow{\Omega},\overrightarrow{R_a},\overrightarrow{R_b}) = S_3(a,b,c), \qquad (16.6)$$

which are functions of Lagrangian coordinates only. Equations (16.4)–(16.6) are a consequence of the momentum equations. Together with continuity Eq. (1.11), they form a system of fluid dynamics equations for an ideal incompressible fluid in a rotating reference frame.

In the case $\overrightarrow{\Omega} = 0$, Eqs. (16.4)–(16.6) take the following form:

$$\overrightarrow{R_{tb}}\,\overrightarrow{R_c} - \overrightarrow{R_{tc}}\,\overrightarrow{R_b} = S_{10}(a,b,c);$$

$$\overrightarrow{R_{tc}}\,\overrightarrow{R_a} - \overrightarrow{R_{ta}}\,\overrightarrow{R_c} = S_{20}(a,b,c);$$

$$\overrightarrow{R_{ta}}\,\overrightarrow{R_b} - \overrightarrow{R_{tb}}\,\overrightarrow{R_a} = S_{30}(a,b,c).$$

The index "0" indicates motion in a non-rotating reference frame. These are expressions for the Cauchy invariants S_{10}, S_{20}, and S_{30} (see Section 1.3.3). Equations (16.4)–(16.6) give a generalization of the Cauchy invariants for motion in a rotating reference frame.

16.1. Gerstner Waves in a Rotating Fluid

16.1.1. *Pollard's solution*

In the f-plane approximation, Pollard indicated the following exact solution of fluid dynamics equations [4]:

$$\begin{cases} X = a - \dfrac{Am}{k}e^{mc}\sin[k(a-Ut)]; \\[2mm] Y = b + f\dfrac{Am}{k^2U}e^{mc}\cos[k(a-Ut)]; \\[2mm] Z = c + Ae^{mc}\cos[k(a-Ut)]. \end{cases} \qquad (16.7)$$

Here, A and m are positive constants, and k and U are the wave number and the wave phase velocity, respectively. Substituting Eq. 16.7 into the continuity Eq. (1.11) yields

$$D_0 = \det \hat{R} = 1 - m^2 A^2 e^{2mc}.$$

The flow region is given by the condition $c \leq c_0 < 0$, so for unambiguity of the maps (16.7) (the determinant D_0 must not take the value zero, it is necessary that the inequality $A \leq 1/m \exp(mc)$ is fulfilled. These requirements guarantees the absence of self-intersections on the wave profile (in the Gerstner wave, the role of the parameter m is played by the wave number).

By substituting Eq. (16.7) into Eqs. (16.4)–(16.6), we check their correctness on the one hand and on the other, we calculate the value of the generalized Cauchy invariants:

$$S_1 = 0; \quad S_2 = m(k^2 - m^2)UA^2 e^{2mc} + \tilde{f}(1 - m^2 A^2 e^{2mc}); \quad S_3 = f.$$

The value of the parameter m is also found from Eq. (16.5):

$$m^2 = \frac{k^4 U^2}{k^2 U^2 - f^2}. \tag{16.8}$$

Thus, only one free parameter A determining the wave amplitude remains in solution (16.7).

Wave oscillations of particles exponentially decay with depth, so the no-leakage condition at the bottom ($c = -\infty$) is fulfilled. To determine the pressure, we substitute Eq. (16.7) into Eq. 16.3 and neglect the centrifugal force. The expression for pressure is

$$p - p_0 = \rho \frac{mgA^2}{2}[e^{2mc} - e^{2mc_0}] - \rho g(c - c_0). \tag{16.9}$$

As for the Gerstner wave, it depends only on the vertical Lagrangian coordinate. When deriving relation (16.9), as a consequence of the independence of pressure on the free surface from time, we obtain the dispersion relation of waves:

$$U^2(k^2 U^2 - f^2) = (g - \tilde{f}U)^2.$$

If there is no rotation (the Coriolis parameters are zero), this equation turns into a dispersion relation for Gerstner waves. The wave runs from west to east, and its crests are parallel to the Y-axis.

It follows from Eqs. (16.7) and (16.8) that fluid particles move in circles:

$$(X - a)^2 + (Y - b)^2 + (Z - c)^2 = \frac{m^2 A^2}{k^2} e^{2mc}.$$

The center of each circle is at the point (a, b, c) not coinciding with the initial position of the particle, and the rotation radius is $mA \exp(2mc)/k$. Comparing the last two expressions of the solution (16.7), we conclude that the motion occurs identically in all planes parallel to the plane

$$Y - f\frac{m}{k^2 U} Z - b + f\frac{m}{k^2 U} c = 0,$$

which forms an angle $\arctan(fm/k^2 U)$ with the Z-axis. The rotation trajectories of the particles rest in these planes. At the equator, $f = 0$ and $m = k$, and Pollard's solution converts to Gerstner's solution (see Eq. (5.4), the coordinate c now plays the role of b). At the equator, the particles oscillate in the XOZ plane, and at $f \neq 0$, the angle of inclination of the plane of their oscillations in each hemisphere is directed toward the corresponding pole. However, as Pollard himself concluded [4], the value of this angle is extremely low.

The stability of Pollard waves has been studied in the short-wave limit. Above a certain threshold of steepness, these waves become unstable with respect to perturbations transverse to the direction of propagation (west−east) [5].

Constantin and Monismith considered the propagation of Pollard waves against the background of a homogeneous zonal flow U_0. If the term $-U_0 t$ is added to the first relation in Eq. (16.7), these expressions will still serve as an exact solution of the rotating fluid equations. But the dispersion relation will be different. Considering this equation, Constantin and Monismith [6] pointed out the possibility of the existence of two types of waves. The first one, described by Pollard, is a somewhat modified Gerstner wave in which there is no drift of fluid particles. The second type (inertial Gerstner wave) is characterized by a slower propagation speed. It is "bound" to the mean flow and cannot occur in its absence.

It is also possible to take into account the effect of the Earth's rotation on the Gerstner edge waves [7]. The form of the exact solution is similar to Eq. (5.11); only the scales of exponential damping

292   *Analytical Fluid Dynamics in Lagrangian Variables*

are now different in the vertical and horizontal directions. The papers by Mollo–Christensen [7, 8] are extremely concise. Their ideas are discussed and developed more fully and in detail for the equatorial region [9] and for an arbitrary latitude [10, 11].

16.1.2. *Nonlinear drift flows generated by wind in Arctic regions*

We would like to draw attention to an example of describing a geophysical flow when Lagrangian and Eulerian approaches are combined, and the analytical solution contains both Eulerian and Lagrangian coordinates. We consider the motion of a viscous fluid in the Arctic regions (only the parameter f will be included in the momentum equations) [12, 13].

We use primes to denote physical (dimensional) variables; these will be removed if we non-dimensionalize them. Consider a Cartesian coordinate system in which the zonal coordinate x' points east, the meridional coordinate y' points north, and the vertical coordinate z' points vertically upward. The governing equations in the f-plane approximation for the Arctic Ocean outside the small neighborhood of the North Pole (which avoids the singularity problem arising from meridian convergence at the pole) are the Navier–Stokes equations [14]

$$\frac{Du'}{Dt} - f'v' = -\frac{1}{\rho'(z)}\frac{\partial P'}{\partial x'} + \mu'\left(\frac{\partial^2 u'}{\partial x'2} + \frac{\partial^2 u'}{\partial y'^2}\right) + \nu'\frac{\partial^2 u'}{\partial z'^2},$$

(16.10)

$$\frac{Dv'}{Dt} + f'u' = -\frac{1}{\rho'(z)}\frac{\partial P'}{\partial y'} + \mu'\left(\frac{\partial^2 v'}{\partial x'^2} + \frac{\partial^2 v'}{\partial y'^2}\right) + \nu'\frac{\partial^2 v'}{\partial z'^2},$$

(16.11)

and

$$\frac{Dw'}{Dt} = -\frac{1}{\rho'(z')}\frac{\partial P'}{\partial z'} - g' + \mu'\left(\frac{\partial^2 w'}{\partial x'^2} + \frac{\partial^2 w'}{\partial y'^2}\right) + \nu'\frac{\partial^2 w'}{\partial z'^2}, \quad (16.12)$$

coupled with the mass conservation equation

$$\frac{D\rho'}{Dt'} + \rho'\left(\frac{\partial u'}{\partial x'} + \frac{\partial v'}{\partial y'} + \frac{\partial w'}{\partial z'}\right) = 0,$$

(16.13)

Exact Solutions for Waves 293

where t' is the time, (u', v', w) is the velocity vector, P' is the pressure, g' is the constant gravity acceleration at the Earth's surface, ρ' is the depth-dependent water density, $f' = 2\Omega'$ is the Coriolis parameter (the constant rate of rotation of the Earth about its polar axis is $\Omega' \approx 7.29 \times 10^{-5}$ rad s^{-1}), μ' and ν' are the constant horizontal and vertical viscosity coefficients, respectively, and

$$\frac{D}{Dt'} = \frac{\partial}{\partial t'} + u'\frac{\partial}{\partial x'} + v'\frac{\partial}{\partial y'} + w'\frac{\partial}{\partial z'}$$

is the material derivative. The boundary conditions related to wind drift solutions of Eqs. (16.10)–(16.13) are the condition of surface pressure

$$P' = P'_{\text{atm}} \quad \text{on } z' = 0,$$

where P'_{atm} is the (constant) atmospheric pressure, the kinematic boundary condition

$$w' = 0 \quad \text{on } z' = 0,$$

and the surface boundary condition, describing the horizontal momentum exchange at the ice-covered surface (see [12] for details).

Let us introduce the characteristic horizontal L' and vertical H' scales of motion. Typical scales for the Arctic flows are $H' = 50$ m and $L' = 10$ km. Defining the dimensionless variables $t, x, y, z, u, v, w, \rho, p, \mu$, and ν by

$$\left.\begin{aligned}
t' &= (L'/U')t, \quad (x', y') = L'(x, y), \quad z' = H'z, \\
(u', v') &= U'(u, v), \quad w' = W'w, \\
\rho' &= \overline{\rho'}\rho, \quad P' = P'_{\text{atm}} + \overline{\rho'}g'H' \int_z^0 \rho(s)ds + \overline{\rho'}U'^2 p, \\
\mu' &= \mu U'H', \quad \nu' = \nu U'H'^2/L'
\end{aligned}\right\}$$

and assuming $\overline{\rho'} = 1030$ kg m^{-3} to be the average density of the Arctic Ocean water, $U' = 0.1$ m s^{-1} and $W' = 10^{-6}$ m s^{-1} to be the horizontal and vertical velocity scales of the ocean flow, respectively, and viscosities $\mu' \approx 50$ m^{2}s^{-1} and $\nu' \approx 0.025$ m^2 s^{-1} to be typical of the averaged vortex, we obtain the non-dimensional form of the governing Eqs. (16.10)–(16.13):

$$\frac{\partial u}{\partial t} + u\frac{\partial u}{\partial x} + v\frac{\partial u}{\partial y} + \frac{\varepsilon}{\delta}w\frac{\partial u}{\partial z} - fv$$

$$= -\frac{1}{\rho}\frac{\partial p}{\partial x} + \delta\mu\left(\frac{\partial^2 u}{\partial x^2} + \frac{\partial^2 u}{\partial y^2}\right) + \nu\frac{\partial^2 u}{\partial z^2}, \tag{16.14}$$

$$\frac{\partial v}{\partial t} + u\frac{\partial v}{\partial x} + v\frac{\partial v}{\partial y} + \frac{\varepsilon}{\delta}w\frac{\partial v}{\partial z} + fu$$

$$= -\frac{1}{\rho}\frac{\partial p}{\partial y} + \delta\mu\left(\frac{\partial^2 v}{\partial x^2} + \frac{\partial^2 v}{\partial y^2}\right) + \nu\frac{\partial^2 v}{\partial z^2}, \tag{16.15}$$

$$\delta\varepsilon\left(\frac{\partial w}{\partial t} + u\frac{\partial w}{\partial x} + v\frac{\partial w}{\partial y} + \frac{\varepsilon}{\delta}w\frac{\partial w}{\partial z}\right)$$

$$= -\frac{1}{\rho}\frac{\partial p}{\partial z} + \varepsilon\delta^2\mu\left(\frac{\partial^2 w}{\partial x^2} + \frac{\partial^2 w}{\partial y^2}\right) + \varepsilon\delta\nu\frac{\partial^2 w}{\partial z^2}, \tag{16.16}$$

and

$$\frac{\partial \rho}{\partial t} + \frac{\partial(\rho u)}{\partial x} + \frac{\partial(\rho v)}{\partial y} + \frac{\varepsilon}{\delta}\frac{\partial(\rho w)}{\partial z} = 0. \tag{16.17}$$

These non-dimensional equations are expressed in terms of the shallowness parameter δ and the ratio ε between the vertical and horizontal velocity scales, given by

$$\delta = H'/L' \quad \text{and} \quad \varepsilon = W'/U',$$

and we denote $f = 2\Omega'L'/U' \approx 145$. Within the mixed layer, we can make use of the incompressibility condition $D\rho'/Dt' = 0$ (see the discussion in other work [12]), whose non-dimensional form

$$\frac{\partial \rho}{\partial t} + u\frac{\partial \rho}{\partial x} + v\frac{\partial \rho}{\partial y} + \frac{\varepsilon}{\delta}w\frac{\partial \rho}{\partial z} = 0$$

simplifies Eq. (16.17) to

$$\frac{\partial u}{\partial x} + \frac{\partial v}{\partial y} + \frac{\varepsilon}{\delta}\frac{\partial w}{\partial z} = 0. \tag{16.18}$$

Exact Solutions for Waves 295

The shallow-water regime $\delta \to 0$, which is adequate for a study of wind drift flows, is characterized by $\delta \to 0$ and $\varepsilon = o(\delta)$ with the dominant viscous term in the vertical direction (to be retained in the leading order). For the specified scales, we have $\delta \approx 10^{-3}$ and $\varepsilon \approx 10^{-5}$, with $\mu = O(1)$ and $\nu = O(1)$. In the limit $\delta \to 0$, we obtain from Eqs. (16.14)–(16.16) and (16.18) the following system for the leading-order dynamics of the wind drift Arctic flows outside the small neighborhood of the North Pole:

$$\frac{\partial u}{\partial t} + u\frac{\partial u}{\partial x} + v\frac{\partial u}{\partial y} - fv = -\frac{1}{\rho}\frac{\partial p}{\partial x} + \nu\frac{\partial^2 u}{\partial z^2}, \qquad (16.19)$$

$$\frac{\partial v}{\partial t} + u\frac{\partial v}{\partial x} + v\frac{\partial v}{\partial y} + fu = -\frac{1}{\rho}\frac{\partial p}{\partial y} + \nu\frac{\partial^2 v}{\partial z^2}, \qquad (16.20)$$

$$0 = \frac{\partial p}{\partial z}, \qquad (16.21)$$

$$\frac{\partial u}{\partial x} + \frac{\partial v}{\partial y} = 0. \qquad (16.22)$$

Using the Lagrangian framework, a study [12] provided an explicit exact solution to Eqs. (16.19)–(16.22) by setting $p \equiv 0$ and specifying, at time t, the complex coordinate of fluid particles

$$X(t; a, b, z) + iY(t; a, b, z)$$
$$= a + ib - d(0)e^{(1+i)\lambda z}t + \frac{1}{k}e^{k(b+z)} \qquad (16.23)$$
$$\exp\left[i\left(\frac{\pi}{2} + k(a - z) - (f + 2\nu k^2)t\right)\right],$$

representing trochoids parametrically (here, $\lambda = \sqrt{f/2\nu}$, $d(0)$ is a constant value and k is the wave number). Let us denote the oscillation period as $T = 2\pi/c_*k$, where $c_* = (f/k) + 2\nu k$. The mean drift flow

$$\frac{1}{T}\int_0^T (u(t; a, b, z) + iv(t; a, b, z))dt = d(0)e^{(1+i)\lambda z} \qquad (16.24)$$

is precisely the classical Ekman spiral (see [14]), its oscillatory perturbation

$$\frac{1}{k}e^{k(b+z)}\exp\left[i\left(\left(\frac{\pi}{2} + k(a - z) - (f + 2\nu k^2)t\right)\right)\right],$$

being almost inertial since its frequency is

$$f + 2\nu k^2 \approx f,$$

given that $\nu = O(1)$ and $k \ll 1$ (the oscillation period T corresponds to about 12 h). The quantity $d(0)$ represents the mean wind drift flow at the surface $z = 0$. Note that the depth-averaged mean drift flow of the solution (16.24), given by

$$\int_{-\infty}^{0} d(0)e^{(1+i)\lambda z}dz = \frac{d(0)}{\lambda\sqrt{2}}\, e^{i\pi/4}$$

due to Eq. (16.24), is 45° to the right of the mean surface flow $d(0)$. Note that the parametric representation of Eq. (16.24) resembles, for a fixed value of z, the trochoidal particle paths in the Gerstner wave (see Section 5.1). We can thus regard the obtained solution to Eq. (16.24) as a nonlinear merger of the Ekman spiral and the Gerstner flow (see Fig. 16.2).

16.2. Waves in the Equatorial Region

Near the equator, in the β-plane approximation, Eqs. (16.4)–(16.6) will be rewritten in the form [3]

$$\frac{D(X_t, X)}{D(b, c)} + \frac{D(Y_t, Y)}{D(b, c)} + \frac{D(Z_t, Z)}{D(b, c)}$$

$$+ 2\Omega\frac{D(Z \cdot X)}{D(b, c)} + \beta Y\frac{D(X, Y)}{D(b, c)} = S_1(a, b, c);$$

$$\frac{D(X_t, X)}{D(c, a)} + \frac{D(Y_t, Y)}{D(c, a)} + \frac{D(Z_t, Z)}{D(c, a)}$$

$$+ 2\Omega\frac{D(Z \cdot X)}{D(c, a)} + \beta Y\frac{D(X, Y)}{D(c, a)} = S_2(a, b, c); \qquad (16.25)$$

$$\frac{D(X_t, X)}{D(a, b)} + \frac{D(Y_t, Y)}{D(a, b)} + \frac{D(Z_t, Z)}{D(a, b)}$$

$$+ 2\Omega\frac{D(Z \cdot X)}{D(a, b)} + \beta Y\frac{D(X, Y)}{D(a, b)} = S_3(a, b, c).$$

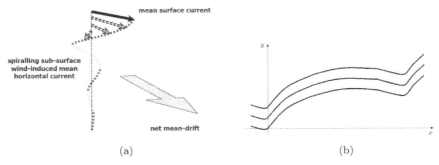

Fig. 16.2. The nonlinear solution (16.24) consists of inertial oscillations superimposed on a mean steady flow that spirals and decays rapidly with depth — the classical Ekman spiral. (a) Depth dependence of the mean horizontal wind drift flow, obtained by averaging the inertial oscillations in the solution. The depth-averaged net flow is 45° to the right of the surface flow. (b) The trochoidal particle paths at a fixed depth level, with inertial horizontal oscillations about the mean flow.

The f-plane approximation at the equator follows from Eqs. (16.25) if we assume $\beta = 0$ (in this case, $f = 0$ and $\tilde{f} = 2\Omega$).

16.2.1. *f-plane approximation*

We will be interested in two-dimensional equatorial flows in planes parallel to the XOZ plane, i.e., $Y = b$. Continuity Eq. (1.11) will take the form

$$\frac{D(X, Z)}{D(a, c)} = D_0(a, c), \qquad (16.26)$$

where D_0 is a time-independent function, and from the equations for Cauchy invariants it follows that

$$\frac{D(X_t, X)}{D(a, c)} + \frac{D(Z_t, Z)}{D(a, c)} = -S_2(a, c) - 2\Omega D_0(a, c). \qquad (16.27)$$

This system of two equations is equivalent to the systems (5.1)–(5.2) with the only difference being that the role of the function Y is now played by Z and the role of the b coordinate by the variable c. Relations (16.26) and (16.27) are a system of equations of two-dimensional fluid dynamics; therefore, all exact solutions for plane waves in a non-rotating fluid will remain valid in the f-plane approximation as well.

298   *Analytical Fluid Dynamics in Lagrangian Variables*

In this case, the expressions for pressure, the vorticity of waves, and their dispersion relations will change. Following similar reasoning, Hsu gave a description of the Gerstner wave in the equatorial region in the f-plane approximation [15] and Kluczek studied the effect of a homogeneous flow on it [16]. In another study [17], the steepness threshold above which the Gerstner wave becomes unstable was found.

The solution of Eqs. (16.26) and (16.27) of the form $X = X(a, c, t)$ and $Z = Z(a, c, t)$ can be generalized by assuming

$$Y = b + \sigma(a, c)t.$$

This corresponds to the superposition of the meridional flow with profile $\sigma(a, c)$ on the known two-dimensional motion. As a result, the flow vorticity and the values of the invariants S_1 and S_3, which take the values $S_1 = -\sigma'_c$ and $S_3 = \sigma'_a$, will change. Henry studied the effect of such flows on the Gerstner wave [18], Kluczek studied their effect on the Gerstner wave in the flow [16], and the author of this book examined their effect on Ptolemaic flows in the equatorial region [2].

Here, let consider the Gerstner wave generated by a traveling harmonic pressure wave. The pressure p on the profile of the Gerstner wave has the form [2]

$$\frac{p - p_0}{\rho} = -gc + \frac{\omega(\omega + 2\Omega)}{2} A^2 e^{2kc}$$

$$+ [\omega(\omega + 2\Omega)k^{-1} - g]A \cos(ka - \omega t). \tag{16.28}$$

Conventionally, the boundary condition of pressure constancy on the free surface ($c = 0$) is set for waves on water. Consequently, by equating the multiplier before the cosine to zero in Eq. (16.28), we obtain the dispersion relation of waves [15]. However, it can be assumed that the pressure distribution in the form of a harmonic traveling wave is maintained on the free surface as a result of wind action

$$p^* = p_1 + p_2 A \cos(ka - \omega t), \tag{16.29}$$

where p_1 and p_2 are constant values satisfying the following relations:

$$p_1 = p_0 + \frac{\omega(\omega + 2\Omega)}{2}\rho A^2; \quad p_2 = \rho[\omega(\omega + 2\Omega)k^{-1} - g]A.$$

$$\tag{16.30}$$

Exact Solutions for Waves   299

Under these conditions, it can be said that Gerstner solution (5.4) corresponds to stationary trochoidal waves on the surface of the fluid supported by external pressure (16.29). By the known ω and k, the wave amplitude A is found from the second equation of this system and the quantity p_0 is found from the first equation. The elevation of the free surface is determined by the formula $Z = A\cos(ka - \omega t)$, so the pressure changes in phase with the profile in the case of positive values of p_2 and antiphase in the case of negative values. The case $p_2 = 0$ corresponds to the Gerstner wave with constant pressure on the profile.

By solving the quadratic Eq. (16.15) with respect to ω, we obtain the dispersion relation for waves

$$\omega = \pm\sqrt{\Omega^2 + \left(g + \frac{p_2}{\rho A}\right)k} - \Omega. \qquad (16.31)$$

We assume $(g + p_2/\rho A) > 0$. Choosing the plus sign, we obtain a wave running eastward, while the minus sign corresponds to its westward propagation. It is interesting to compare dispersion relation (16.31) with the dispersion relation for equatorial waves in a homogeneous flow [16]. The value of $p_2/2\rho A\Omega$ is similar to the flow velocity. Thus, the wind either accelerates or decelerates the waves depending on the sign of p_2.

The considered solution is also interesting from the point of view of the implementation of Gerstner waves in laboratory and field conditions. One of the possible mechanisms of their generation can be the wind, as follows from our analysis.

16.2.2. *Trapped waves (β-plane approximation)*

Constantin found the exact solution to the system (16.25):

$$\begin{cases} X = a - \dfrac{1}{k}e^{k[c-h(b)]}\sin k(a - Ut); \\[2mm] \qquad\qquad Y = b; \\[2mm] Z = c + \dfrac{1}{k}e^{k[c-h(b)]}\cos k(a - Ut), \end{cases} \qquad (16.32)$$

where $h(b) = \frac{\beta}{2(kU+2\Omega)}b^2$, and the phase velocity of the wave is

$$U = \frac{\sqrt{\Omega^2 + kg} - \Omega}{k}.$$

300   *Analytical Fluid Dynamics in Lagrangian Variables*

The Lagrangian variables a and b vary within $(-\infty, \infty)$, whereas c lies within $(-\infty, c_0)$, where $c_0 < 0$. Equation (16.32) describes equatorial surface waves traveling with velocity U in the easterly direction [19]. These are periodic spatial waves whose amplitude decreases exponentially in the meridional direction. As a result, they are referred to as trapped. When $h = 0$, Eq. (16.32) converts into a Gerstner solution. An additional exponentially decaying multiplier for the amplitude is the "highlight" of this exact solution.

Expressions for generalized Cauchy invariants of waves (16.32) have the following form [3]:

$$S_1 = 0, \, S_2 = 2\Omega - 2(kU + \Omega)e^{2\varsigma};$$

$$S_3 = \beta b \left[1 - \frac{2(kU + \Omega)}{kU + 2\Omega} e^{2\varsigma} \right], \qquad (16.33)$$

$$\varsigma = k[c - h(b)].$$

The zonal component of the vector $\overrightarrow{S}\{S_1, S_2, S_3\}$ is zero. The vorticity $\overrightarrow{\omega}$ for waves (16.32) is determined by the following equalities [19]:

$$\overrightarrow{\omega} = D_0^{-1}\left\{ -bkU^2 g^{-1}\beta e^{\varsigma}\sin\theta, \right.$$

$$\left. - 2kUe^{2\varsigma}, \quad bkU^2 g^{-1}\beta(e^{\varsigma}\cos\theta - e^{2\varsigma}) \right\};$$

$$D_0^{-1} = 1 - e^{2\varsigma}; \quad \theta = k(a - Ut). \qquad (16.34)$$

All three of its components are non-zero, and the zonal and vertical components depend on time. A comparison of Eqs. (16.33) and (16.34) gives a clear difference between the vorticity vector and the vector of Lagrangian invariants.

The pressure in the fluid is written as follows:

$$p - p_0 = \rho g \left(\frac{e^{2\varsigma}}{2k} - c \right) - \rho g \left(\frac{e^{2kc_0}}{2k} - c_0 \right).$$

The form of the free surface at the equator, where $b = 0$, is obtained from Eq. (16.32) by assuming $c = c_0$. Qualitatively, trapped equatorial waves can be compared with edge waves along a vertical coastal wall at the equator.

Like their simpler counterparts, Gerstner waves and Gerstner-type edge waves, trapped equatorial waves are linearly unstable when their steepness exceeds a certain limit [20].

Expression (16.32) is a unique example of an exact solution of the equations of fluid motion in the equatorial region. These one are generalized to the case of homogeneous zonal near-surface flow [21–23], account for the action of centrifugal forces [24], and correct for the gravity force within the framework of the standard β-plane model [25] and edge waves on a sloping beach lying parallel to the equator [26]. In other works [27, 28], Eq. (16.32) is used to describe waves propagating against the background of a zonal flow along an arbitrary latitude. In this case, an unconventional β-approximation is employed, when the parameter f varies with latitude and the parameter $\tilde{f}$ is assumed constant. This approximation for the spatial problem is very rough, and the obtained solutions for waves cannot be called exact.

16.2.3. *Accounting for stratification*

The motion of a multilayer fluid with trapped Gerstner-like waves in the equatorial region was studied by Constantin [29, 30]. The role of Gerstner waves in his works is played by the equatorial waves he discovered [19]. In one of Constantin's studies [29], equatorially trapped waves propagating eastward in a thermocline (symmetric about the equator) are considered. The fluid above the thermocline is assumed to be at rest (similar to the Mollo–Christensen two-layer model). In another of his studies [30], a variant of "anti-Gerstner" waves was explored when the amplitude of wave disturbances decreases above the thermocline and a homogeneous flow moves under it. The stability of these two-layer models is studied [31–33] and their generalizations to the case of additional background flows are discussed [34, 35]. Kluczek, following another example [30], constructed a solution for the internal Pollard wave [36, 37], in our terminology, the "anti-Pollard" wave.

Weber, studying the properties of weakly nonlinear waves at the boundary of two fluids [38, 39], suggests that any wave for which there is no drift of fluid particles should be called a Gerstner-like wave. Obviously, all the exact solutions analyzed and mentioned by us have this property. In this sense, it probably makes sense to talk about a family of different waves, which can be united by the name Gerstner and called Gerstner-like.

References

[1] Pedlosky, J. (1979). *Geophysical Fluid Dynamics* (Springer, New York).

[2] Abrashkin, A. (2019). Wind generated equatorial Gerstner-type waves. *Discrete Contin. Dyn. Syst.*, 39(8), pp. 4443–4453.

[3] Abrashkin, A. (2019). Generalization of Cauchy invariants for equatorial β-plane flows. *Deep-Sea Research Part II*, 160, pp. 3–6.

[4] Pollard, R.T. (1970). Surface waves with rotation: An exact solution. *J. Geophys. Res.*, 75, pp. 5895–5898.

[5] Ionescu-Kruse, D. (2016). Instability of Pollard's exact solution for geophysical ocean flows. *Phys. Fluids*, 28, 086601.

[6] Constantin, A. and Monismith, S. G. (2017). Gerstner waves in the presence of mean currents and rotation. *J. Fluid Mech.*, 820, pp. 511–528.

[7] Mollo-Christensen, E. (1978). Edge waves in a rotating stratified fluid, an exact solution. *J. Phys. Ocean*, 9, pp. 226–229.

[8] Mollo-Christensen, E. (1982). Allowable discontinuities in a Gerstner wave. *Phys. Fluids*, 25, pp. 586–587.

[9] Matioc, A.-V. (2012). An exact solution for geophysical equatorial edge waves over a sloping beach. *J. Phys. A: Math. Theor.*, 45, 365501.

[10] Weber, J.E.H. (2012). A note on trapped Gerstner waves. *J. Geoph. Res.*, 117, C03048.

[11] Ionescu-Kruse, D. (2015). An exact solution for geophysical edge waves in the f-plane approximation, *Nonl. Anal. Real World Appl.*, 24, pp. 190–195.

[12] Constantin, A. (2022). Nonlinear wind-drift ocean currents in arctic regions. *Geophys. Astrophys. Fluid Dyn.*, 116(2), pp. 101–115.

[13] Constantin, A. (2022). Comments on: nonlinear wind-drift ocean currents in arctic regions. *Geophys. Astrophys. Fluid Dyn.*, 116(2), pp. 116–121.

[14] Talley, L.D., Pickard, G.L., Emery, W.J. and Swift, J.H. (2011). *Descriptive Physical Oceanography: An Introduction* (Elsevier, Amsterdam).

[15] Hsu, H.-C. (2015). An exact solution for equatorial waves. *Monatsh. Math.*, 176, pp. 143–152.

[16] Kluczek, M. (2018). Equatorial water waves with underlying currents in the f-plane approximation. *Appl. Anal.*, 97, pp. 1867–1880.

[17]   Henry, D. and Hsu, H.-C. (2015). Instability of equatorial water waves in the f-plane, *Discrete Contin. Dyn. Syst.*, 35, pp. 909–916.

[18]   Henry, D. (2016). Exact equatorial water waves in the f-plane. *Nonl. Anal. Real World Appl.*, 28, pp. 284–289.

[19]   Constantin, A. (2012). An exact solution for equatorially trapped waves. *J. Geophys. Res.*, 117, C05029.

[20]   Constantin, A. and Germain, P. (2013). Instability of some equatorially trapped waves. *J. Geophys. Res.-Oceans*, 118, pp. 2802–2810.

[21]   Henry, D. (2013). An exact solution for equatorial geophysical water waves with an underlying current, *Eur. J. Mech. (B/Fluids)*, 38, pp. 18–21.

[22]   Henry, D. and Sastre-Gomez, S. (2016). Mean flow velocities and mass transport for equatorially-trapped water waves with an underlying current. *J. Math. Fluid Mech.*, 18, pp. 795–804.

[23]   Genoud, F. and Henry, D. (2014). Instability of equatorial water waves with an underlying current. *J. Math. Fluid Mech.*, 16, pp. 661–667.

[24]   Henry, D. (2016). Equatorially trapped nonlinear water waves in a β-plane approximation with centripetal forces. *J. Fluid Mech.*, 804, R1–11.

[25]   Henry, D. (2017). On three-dimensional Gerstner-like equatorial water waves, *Philos. Trans. R. Soc. A*, 376, 20170088.

[26]   Ionescu-Kruse, D. (2015). An exact solution for geophysical edge waves in the β-plane approximation. *J. Math. Fluid Mech.*, 17, pp. 699–706.

[27]   Chu, J., Ionescu-Kruse, D. and Yang, Y. (2019). Exact solutions and instability for geophysical waves at arbitrary latitude. *Discrete Cont. Dyn. Syst.*, 39(8), pp. 4399–4414.

[28]   Chu, J., Ionescu-Kruse, D. and Yang, Y. (2019). Exact solution and instability for geophysical waves with centripetal forces and at arbitrary latitude. *J. Math. Fluid Mech.*, 21, 19.

[29]   Constantin, A. (2013). Some three-dimensional nonlinear equatorial flows. *J. Phys. Oceanogr.*, 43, pp. 165–175.

[30]   Constantin, A. (2014). Some nonlinear, equatorially trapped, non-hidrostatic internal geophysical waves. *J. Phys. Oceanogr.*, 44, pp. 781–789.

[31]   Henry, D. and. Hsu, H.-C. (2015). Instability of internal equatorial water waves. *J. Diff. Eq.*, 258, pp. 1015–1024.

[32]   Ionescu-Kruse, D. (2016). Instability of equatorially trapped waves in stratified water. *Ann. Mat. Pura Appl*, 195, pp. 585–599.

[33]   Rodrǵuez-Sanjurjo, A. (2019). Instability of equatorially-trapped non-hydrostatic internal geophysical water waves. *Wave Motion*, 88, pp. 144–152.

[34] Kluczek, M. (2017). Exact and explicit internal equatorially-trapped water waves with underlying currents. *J. Math. Fluid Mech.*, 19, pp. 305–314.

[35] Rodrǵuez-Sanjurjo, A. and Kluczek, M. (2016). Mean flow properties for equatorially trapped internal water wave–current interactions. *Appl. Anal.*, 96, pp. 2333–2345.

[36] Kluczek, M. (2019). Exact Pollard-like internal water waves. *J. Nonl. Math. Phys.*, 26(1), pp. 133–146.

[37] Kluczek, M. (2018). Physical flow properties for Pollard-like internal water waves. *J. Math. Phys.*, 59, 123102.

[38] Weber, J.E.H. (2018). An interfacial Gerstner-type trapped wave. *Wave Motion*, 77, pp. 186–194.

[39] Weber, J.E.H. (2019). A Lagrangian study of internal Gerstner- and Stokes type gravity waves. *Wave Motion*, 88, pp. 257–264.

Chapter 17

Vortices in a Rotating Fluid

Abstract

In this chapter, we study nonlinear spatial non-stationary vortex structures. However, some more general versions of standard approximations will be used for their analytical description within the framework of complete equations of fluid dynamics of an ideal incompressible fluid.

17.1. Motion of Unsteady Columns in an Exponentially Stratified Fluid in the Boussinesq Approximation

We will study the motion of a fluid in a uniformly rotating frame of reference [1]. The flow is described by continuity Eq. (1.11) and momentum Eq. (16.1). Let the rotational angular velocity vector be directed exactly along the z-axis. For oceanic problems, this means that high-latitude regions near the poles are being considered. We also assume that the fluid is exponentially stratified vertically (along the Lagrangian coordinate c), so that the Brunt−Väisälä frequency $N^2 = -g\rho_c/\rho = \text{const}$.

Similar to the transformations at the beginning of Chapter 16, we pass to Lagrangian variables in Eq. (16.2) and exclude pressure. The three equations, equivalent to the momentum equations, can be written as follows [1]:

$$\frac{\partial}{\partial t}[\vec{R}_{ta}\vec{R}_b - \vec{R}_{tb}\vec{R}_a + 2\,(\overrightarrow{\Omega}\,\vec{R}_a\vec{R}_b)] = 0, \qquad (17.1)$$

$$\frac{\partial}{\partial t}[\vec{R}_{tc}\vec{R}_a - \vec{R}_{ta}\vec{R}_c + 2(\vec{\Omega}\vec{R}_a\vec{R}_c)] = N^2 Z_a, \qquad (17.2)$$

$$\frac{\partial}{\partial t}[\vec{R}_{tc}\vec{R}_b - \vec{R}_{tb}\vec{R}_c + 2(\vec{\Omega}\vec{R}_c\vec{R}_b)] = N^2 Z_b. \qquad (17.3)$$

When writing Eqs. (17.1) and (17.2), it was assumed that the acceleration of fluid particles in a rotating system, Coriolis acceleration, and centripetal acceleration are low compared to free-fall acceleration g. The authors of another study [1] called it the generalized Boussinesq approximation.

Thus, we have the continuity equation

$$\frac{\partial}{\partial t}\frac{D(X,Y,Z)}{D(a,b,c)} = 0 \qquad (17.4)$$

and three momentum Eqs. (17.1)–(17.3). Assume that for unknown functions the following relations are fulfilled:

$$X_c = 0, \quad Y_c = 0, \quad Z = c + Z_*(a,b,t). \qquad (17.5)$$

In this case, Eq. (17.4) is reduced to a simpler condition:

$$\frac{\partial}{\partial t}(X_a Y_b - X_b Y_a) = 0. \qquad (17.6)$$

It follows that $\frac{\partial}{\partial t}(\vec{\Omega}\vec{R}_a\vec{R}_b) = 0$; therefore, Eq. (17.1) takes the form

$$\frac{\partial}{\partial t}(X_{ta}X_b - X_{tb}X_a + Y_{ta}Y_b - Y_{tb}Y_a + Z_{*ta}Z_{*b} - Z_{*tb}Z_{*a}) = 0.$$
$$(17.7)$$

With allowance for Eq. (17.5), Eqs. (17.2) and (17.3) will be rewritten as follows:

$$Z_{*tta} + N^2 Z_{*a} = 0, \qquad (17.8)$$

$$Z_{*ttb} + N^2 Z_{*b} = 0. \qquad (17.9)$$

We note at once that the solution does not depend on the angular velocity.

Equations (17.8) and (17.9) are elementary integrated. The general form of Z_* is as follows:

$$
\begin{aligned}
Z_* &= C_1(a,b)\sin(Nt) + C_2(a,b)\cos(Nt) \\
&= A(a,b)\sin(Nt + \theta(a,b)).
\end{aligned}
\tag{17.10}
$$

The fluid particles make vertical harmonic oscillations with frequency N. If $N = 0$ (no stratification), then

$$
Z_* = C_3(a,b)t.
\tag{17.11}
$$

In this case, the fluid particles move vertically and their velocity profile is $C_3(a,b)$.

By virtue of relations (17.8) and (17.9), Eq. (17.7) will also be simplified, i.e., the dependence on Z_* disappears in it, and it takes the following form:

$$
\frac{\partial}{\partial t}(X_{ta}X_b - X_{tb}X_a + Y_{ta}Y_b - Y_{tb}Y_a) = 0.
\tag{17.12}
$$

Equations (17.6) and (17.12) coincide with the equations of two-dimensional fluid dynamics (cf. Eqs. (5.1) and (5.2)). The motion in the horizontal plane and in the vertical direction has been separated. One can choose any two-dimensional solution and apply the flows (17.10) or (17.11) to it. The possibility of such a method for homogeneous flows has already been noted by us in Section 9.3 and 16.2.1. The authors of the previously mentioned work [1] constructed original examples by choosing Ptolemaic (Section 8.1) and elliptical (Section 8.3.1) vortices as two-dimensional solutions.

17.2. Lagrangian Equations in Spherical Coordinates

We introduce a set of (right-handed) spherical coordinates (R, Θ, Φ), where R is the distance (radius) from the center of a sphere, Θ (with $0 \leq \Theta \leq \pi$) is the polar angle, and Φ (with $0 \leq \Phi < 2\pi$) is the azimuthal angle.

The Cartesian coordinates $X, Y,$ and Z have the following form:

$$
X = R\sin\Theta\cos\Phi, \quad Y = R\sin\Theta\sin\Phi, \quad Z = R\cos\Phi.
$$

308   *Analytical Fluid Dynamics in Lagrangian Variables*

The Cartesian (a, b, c) and spherical (r, θ, φ) Lagrangian variables are related by similar formulas:

$$a = r \sin \theta \cos \varphi, \quad b = r \sin \theta \sin \varphi, \quad c = r \cos \theta,$$

with the latter having the same meaning as (R, Θ, Φ).

The continuity equation is obtained using the Jacobian multiplication rule

$$\frac{D(X, Y, Z)}{D(a, b, c)} = \frac{D(X, Y, Z)}{D(R, \Theta, \Phi)} \frac{D(R, \Theta, \Phi)}{D(r, \theta, \varphi)} \frac{D(r, \theta, \varphi)}{D(a, b, c)} = D_0(r, \theta, \varphi).$$

Taking into account the equalities

$$\frac{D(X, Y, Z)}{D(R, \Theta, \Phi)} = R^2 \sin \Theta; \quad \frac{D(r, \theta, \varphi)}{D(a, b, c)} = (r^2 \sin \theta)^{-1},$$

we obtain

$$R^2 \sin \Theta \frac{D(R, \Theta, \Phi)}{D(r, \theta, \varphi)} = r^2 D_0 \sin \theta = D_0^*(r, \theta, \varphi). \tag{17.13}$$

This expression represents the continuity equation in spherical Lagrangian coordinates.

Let us now return to the derivation of the momentum equations. We write Eq. (16.3) in expanded form

$$\vec{R}_{tt} \vec{R}_a + 2(\vec{\Omega} \vec{R}_t \vec{R}_a) = -H_a, \tag{17.14}$$

$$\vec{R}_{tt} \vec{R}_b + 2(\vec{\Omega} \vec{R}_t \vec{R}_b) = -H_b, \tag{17.15}$$

$$\vec{R}_{tt} \vec{R}_c + 2(\vec{\Omega} \vec{R}_t \vec{R}_c) = -H_c. \tag{17.16}$$

Using the Jacobian transformation rules, we obtain the partial derivatives of an arbitrary function Q with respect to the spherical Lagrangian variable r:

$$\frac{\partial Q}{\partial r} = \frac{D(Q, \theta, \varphi)}{D(r, \theta, \varphi)} = \frac{D(Q, \theta, \varphi)}{D(a, b, c)} \frac{D(a, b, c)}{D(r, \theta, \varphi)}$$

$$= (r^2 \sin \theta)^{-1} \left[\frac{\partial Q}{\partial a} \frac{D(\theta, \varphi)}{D(b, c)} + \frac{\partial Q}{\partial b} \frac{D(\theta, \varphi)}{D(c, a)} + \frac{\partial Q}{\partial c} \frac{D(\theta, \varphi)}{D(a, b)} \right].$$

Vortices in a Rotating Fluid 309

The derivative with respect to the corresponding Lagrangian coordinate is linear in each term of Eqs. (17.14)–(17.16). By multiplying the equations of this system by $\frac{D(\theta,\varphi)}{D(b,c)}$, $\frac{D(\theta,\varphi)}{D(c,a)}$, and $\frac{D(\theta,\varphi)}{D(a,b)}$, respectively, and adding them, we obtain

$$\vec{R}_{tt}\vec{R}_r + 2(\overrightarrow{\Omega}\,\vec{R}_t\vec{R}_r) = -H_r. \tag{17.17}$$

Two other equations could be obtained in a similar manner:

$$\vec{R}_{tt}\vec{R}_\theta + 2(\overrightarrow{\Omega}\,\vec{R}_t\vec{R}_\theta) = -H_\theta, \tag{17.18}$$

$$\vec{R}_{tt}\vec{R}_\varphi + 2(\overrightarrow{\Omega}\,\vec{R}_t\vec{R}_\varphi) = -H_\varphi. \tag{17.19}$$

The momentum equations have a similar form in Cartesian and spherical coordinates. This is due to the fact that Lagrangian coordinates serve as labels for fluid particles and can arbitrarily be chosen. The constraint is the condition of one-to-one correspondence between labels of particles and their coordinates (the Jacobian of the transformation is not zero). Equations (17.17)–(17.19) have not been quoted so far. In a monograph [2], the left-hand side of the momentum equation remains unchanged, but the gradient of the function H is expressed through derivatives with respect to the Lagrangian variables.

Equations (17.17)–(17.19) together with Eq. (17.13) form complete simultaneous equations of fluid dynamics of an ideal incompressible fluid in spherical Lagrangian coordinates. In vector form, the momentum equations will be written quite compactly. But their representation through unknown functions R, Θ, and Φ looks extremely cumbersome. Therefore, we will not give their full formulation here. However, Eqs. (17.17)–(17.19) will be used to determine the pressure of specific vortex flows further on.

It is more convenient to use Eqs. (16.4)–(16.6) instead of momentum equations to study the flow features since these equations are of the same form. In each of them, differentiation is performed on a certain pair of Lagrangian variables. In terms of unknown functions R, Θ, and Φ, Eqs. (16.4)–(16.6) are written as follows:

$$\frac{D(R_t, R)}{D(b, c)} + R^2\frac{D(\Theta_t, \Theta)}{D(b, c)} + R^2\sin^2\Theta\frac{D(\Phi_t, \Phi)}{D(b, c)} + 2R\Theta_t\frac{D(R, \Theta)}{D(b, c)}$$

$$+ 2\Omega \left(R\sin^2\Theta\frac{D(R,\Phi)}{D(b,c)} + R^2\sin\Theta\cos\Theta\frac{D(\Theta,\Phi)}{D(b,c)} \right) = S_1,$$

$$(17.20)$$

$$\frac{D(R_t,R)}{D(c,a)} + R^2\frac{D(\Theta_t,\Theta)}{D(c,a)} + R^2\sin^2\Theta\frac{D(\Phi_t,\Phi)}{D(c,a)} + 2R\Theta_t\frac{D(R,\Theta)}{D(c,a)}$$

$$+ 2\Omega \left(R\sin^2\Theta\frac{D(R,\Phi)}{D(c,a)} + R^2\sin\Theta\cos\Theta\frac{D(\Theta,\Phi)}{D(c,a)} \right) = S_2,$$

$$(17.21)$$

$$\frac{D(R_t,R)}{D(a,b)} + R^2\frac{D(\Theta_t,\Theta)}{D(a,b)} + R^2\sin^2\Theta\frac{D(\Phi_t,\Phi)}{D(a,b)} + 2R\Theta_t\frac{D(R,\Theta)}{D(a,b)}$$

$$+ 2\Omega(R\sin^2\Theta\frac{D(R,\Phi)}{D(a,b)} + R^2\sin\Theta\cos\Theta\frac{D(\Theta,\Phi)}{D(a,b)}) = S_3.$$

$$(17.22)$$

The vector $\vec{\Omega}$ is considered to be directed along the Z-axis here.

Finally, we pass from a pair of Cartesian Lagrangian coordinates to a pair of spherical Lagrangian variables. Let us discuss the transition process for a simple example and assume that the following simultaneous equations are given:

$$\frac{D(A,B)}{D(b,c)} = C_1, \quad \frac{D(A,B)}{D(c,a)} = C_2, \quad \frac{D(A,B)}{D(a,b)} = C_3, \qquad (17.23)$$

where the right-hand sides are the Lagrangian invariants. Here, we need to find the form of Eq. (17.23) in spherical coordinates r, θ, and φ.

So, we write the first equation of this system as follows:

$$\frac{D(A,B)}{D(b,c)} = \frac{D(A,B,a)}{D(b,c,a)} = \frac{D(A,B,a)}{D(a,b,c)}$$

$$= \frac{D(A,B,a)}{D(r,\theta,\varphi)}\frac{D(r,\theta,\varphi)}{D(a,b,c)} = C_1,$$

or, in expanded form,

$$\frac{D(A,B)}{D(\theta,\varphi)}a_r + \frac{D(A,B)}{D(\varphi,r)}a_\theta + \frac{D(A,B)}{D(r,\theta)}a_\varphi = C_1 r^2\sin\theta. \qquad (17.24)$$

Similarly, one can obtain

$$\frac{D(A,B)}{D(\theta,\varphi)}b_r + \frac{D(A,B)}{D(\varphi,r)}b_\theta + \frac{D(A,B)}{D(r,\theta)}b_\varphi = C_2 r^2 \sin\theta, \quad (17.25)$$

$$\frac{D(A,B)}{D(\theta,\varphi)}c_r + \frac{D(A,B)}{D(\varphi,r)}c_\theta + \frac{D(A,B)}{D(r,\theta)}c_\varphi = C_3 r^2 \sin\theta. \quad (17.26)$$

Expressions (17.24)–(17.26) are expansions of the right-hand sides by elements of one of the rows of the determinant:

$$\frac{D(a,b,c)}{D(r,\theta,\varphi)} = \begin{vmatrix} a_r & a_\theta & a_\varphi \\ b_r & b_\theta & b_\varphi \\ c_r & c_\theta & c_\varphi \end{vmatrix}. \quad (17.27)$$

Multiplying the relations (17.24)–(17.26) by the corresponding minors of the determinant (17.27), we obtain

$$\frac{D(A\cdot B)}{D(\theta,\varphi)} = r^2 \sin^2\theta(C_1\cos\varphi + C_2\sin\varphi + C_3) = C_1^*(r,\theta,\varphi);$$

$$\frac{D(A,B)}{D(\varphi,r)} = r\sin\theta[(C_1\cos\varphi + C_2\sin\varphi) - C_3\sin\theta] = C_2^*(r,\theta,\varphi);$$

$$\frac{D(A,B)}{D(r,\theta)} = (C_2 - C_1)r\sin\varphi = C_3^*(r,\theta,\varphi).$$

Equations (17.20)–(17.22) are a linear combination of Jacobians of the same type. Therefore, we can apply the proven rule to each of them when replacing pairs of Cartesian coordinates with pairs of spherical coordinates. The Lagrangian equations in spherical coordinates could be written as follows [3]:

$$\frac{D(R_t,R)}{D(\theta,\varphi)} + R^2\frac{D(\Theta_t,\Theta)}{D(\theta,\varphi)} + R^2\sin^2\Theta\frac{D(\Phi_t,\Phi)}{D(\theta,\varphi)} + 2R\Theta_t\frac{D(R,\Theta)}{D(\theta,\varphi)}$$

$$+ 2\Omega\left(R\sin^2\Theta\frac{D(R,\Phi)}{D(\theta,\varphi)} + R^2\sin\Theta\cos\Theta\frac{D(\Theta,\Phi)}{D(\theta,\varphi)}\right) = S_1^*(r,\theta,\varphi);$$

$$(17.28)$$

$$\frac{D(R_t, R)}{D(\varphi, r)} + R^2\frac{D(\Theta_t, \Theta)}{D(\varphi, r)} + R^2\sin^2\Theta\frac{D(\Phi_t, \Phi)}{D(\varphi, r)} + 2R\Theta_t\frac{D(R, \Theta)}{D(\varphi, r)}$$

$$+ 2\Omega\left(R\sin^2\Theta\frac{D(R, \Phi)}{D(\varphi, r)} + R^2\sin\Theta\cos\Theta\frac{D(\Theta, \Phi)}{D(\varphi, r)}\right) = S_2^*(r, \theta, \varphi);$$

$$(17.29)$$

$$\frac{D(R_t, R)}{D(r, \theta)} + R^2\frac{D(\Theta_t, \Theta)}{D(r, \theta)} + R^2\sin^2\Theta\frac{D(\Phi_t, \Phi)}{D(r, \theta)} + 2R\Theta_t\frac{D(R, \Theta)}{D(r, \theta)}$$

$$+ 2\Omega\left(R\sin^2\Theta\frac{D(R, \Phi)}{D(r, \theta)} + R^2\sin\Theta\cos\Theta\frac{D(\Theta, \Phi)}{D(r, \theta)}\right) = S_3^*(r, \theta, \varphi).$$

$$(17.30)$$

Here, S_1^*, S_2^*, and S_3^* are invariants depending on spherical Lagrangian coordinates.

17.3. Problem Statement and Approximations Used

We are interested in the details of fluid dynamics in oceanic eddies. For this purpose, we will use spherical coordinates, where R refers to distances from the Earth's center, Θ (and θ) are polar angles, and Φ (and φ) refers to azimuthal angles, i.e., angles of longitude. The values $\pi/2 - \Theta$ and $\pi/2 - \theta$ are conventional angles of latitude, so that the North Pole corresponds to $\Theta, \theta = 0$, the South Pole to $\Theta, \theta = \pi$, and the Equator to $\Theta, \theta = \pi/2$. The vector $\overrightarrow{\Omega}$ is directed from south to north; $\Omega \approx 7.29 \times 10^{-5}\,\mathrm{rad}\,s^{-1}$ is the constant rotational velocity of the Earth. The gravity potential is $\Phi_* = g(R - R_0)$, where g is the acceleration due to gravity.

17.3.1. Thin layer approximation

Let us assume that the Earth is a solid sphere of radius R_0 covered with a water layer of thickness h. The variable R is equal to $R = R_0 + r$, where $0 \leq r \leq h$. In the zero order with respect to the small parameter h/R_0 in Eqs. (17.28)–(17.30), R can be replaced by R_0 (formally, $h = 0$).

Thin layer approximations are characteristic in that the velocity component across the layer is small compared with the velocities

Vortices in a Rotating Fluid

along the layer. We will neglect the vertical motion of the fluid. Fluid particles should move along the spherical surface $r = \text{const}$. The flows on spherical surfaces will be similar, and the no-leakage condition is automatically satisfied on a solid surface $r = 0$.

Suppose that the trajectory coordinates Θ and Φ do not depend on the variable r. So, Eqs. (17.28)–(17.30) could be rewritten as

$$\sin \frac{D(\Theta, \Phi)}{D(\theta, \varphi)} = R_0^{-2} D_0^*(\theta, \varphi), \tag{17.31}$$

$$\frac{D(\Theta_t, \Theta)}{D(\theta, \varphi)} + \sin^2 \Theta \frac{D(\Phi_t, \Phi)}{D(\theta, \varphi)} + 2\Omega \sin \Theta \cos \Theta \frac{D(\Theta, \Phi)}{D(\theta, \varphi)}$$
$$= R^{-2}{}_0 S_1^*(\theta, \varphi). \tag{17.32}$$

The left-hand side of Eq. (17.31) is entirely included in one of the terms of Eq. (17.32). This fact will be used in our further calculations.

Equations (17.31) and (17.32) do not include pressure. It follows from Eq. (17.17) that

$$H = \frac{p}{\rho} + gr - \frac{1}{2}\Omega^2 (R_0 + r)^2 \sin^2 \Theta = H(\theta, \varphi), \tag{17.33}$$

where $H(\theta, \varphi)$ is the function to be defined. The condition

$$p|_{r=h} = P_s(\theta, \varphi)$$

should be fulfilled on the free surface, where $P_s(\theta, \varphi)$ is the pressure inducing the vertical motion under study. This is non-uniform air pressure on the water surface. The function $P_s(\theta, \varphi)$ is determined by Eqs. (17.18) and (17.19) through the known solution of the set of Eqs. (17.31) and (17.32). The function $H(\theta, \varphi)$ is related to it as follows:

$$H(\theta, \varphi) = \frac{P_s(\theta, \varphi)}{\rho} + gh - \frac{1}{2}\Omega^2 (R_0^2 + 2R_0 h) \sin^2 \Theta. \tag{17.34}$$

The flow velocity in this approximation $\vec{V}(0, R_0 \Theta_t, R_0 \Phi_t \sin \Theta)$ has only two components, namely, meridional and azimuthal. Since both of them do not depend on the radius, only the radial component of the vorticity will be non-zero, and therefore it can be written in the form

$$\begin{aligned}
\omega_R &= \frac{1}{R_0 \sin \Theta} \left[\frac{\partial (V_\Phi \sin \Theta)}{\partial \Theta} - \frac{\partial V_\Theta}{\partial \Phi} \right] \\
&= \frac{1}{\sin \Theta} \left[\frac{D(\Phi_t \sin^2 \Theta, \Phi)}{D(\Theta, \Phi)} + \frac{D(\Theta_t, \Theta)}{D(\Theta, \Phi)} \right] \\
&= \frac{R_0^2}{D_0^*} \left[\frac{D(\Theta_t, \Theta)}{D(\theta, \varphi)} + \sin^2 \Theta \frac{D(\Phi_t, \Phi)}{D(\theta, \varphi)} + 2\Phi_t \sin \Theta \cos \Theta \frac{D(\Theta, \Phi)}{D(\theta, \varphi)} \right].
\end{aligned}$$
(17.35)

The formulation of the final expression was based on Eq. (17.31). The vorticity formula (17.35) is similar to the left-hand side of Eq. (17.32), which is the Cauchy integral in the thin layer approximation. The variable Ω was replaced by the function Φ_t in the third term in square brackets of Eq. (17.35). By comparing Eqs. (17.31), (17.32), and (17.35), one can obtain a relation for the vorticity through this function:

$$\omega_R + (2\Omega - \Phi_t) \cos \Theta = \frac{S_1^*}{D_0^*}.$$
(17.36)

The vorticity of fluid particles is not preserved in the general case. It forms the integral of motion when summing the "spin vorticity" [4] with a term that takes into account the azimuthal motion of a fluid.

17.3.2. *Averaged latitude approximation*

Let us introduce some new unknown functions

$$X = \lambda \Phi, \quad \Theta^* = \frac{\pi}{2} - \Theta, \quad Y = \cos \Theta = \sin \Theta^*,$$
(17.37)

and the variable

$$\theta^* = \frac{\pi}{2} - \theta,$$

where λ is the constant, and Θ^* and θ^* are latitudes in the Eulerian and Lagrangian descriptions (both are positive in the northern

Vortices in a Rotating Fluid

hemisphere and negative in the southern hemisphere). Simultaneous Eqs. (17.31) and (17.32) can be reformulated as follows:

$$\frac{D(X,Y)}{D(\varphi,\theta^*)} = \lambda R_0^{-2} D_0^*,$$

(17.38)

$$\lambda^{-2} \cos^2 \theta^* \frac{D(X_t, X)}{D(\varphi, \theta^*)} + \cos^{-2} \theta^* \frac{D(Y_t, Y)}{D(\varphi, \theta^*)} + 2\Omega R_0^{-2} S_0^* \sin \theta^*$$
$$= R_0^{-2} S_1^*.$$

(17.39)

The proposed substitution of variables was fulfilled within the Jacobians only (the reason will be clarified in the following). The set of Eqs. (17.38) and (17.39) is extremely difficult to solve. For simplicity, we propose to replace it with another nonlinear system with possible exact solutions.

The fluid particles in eddies moved around a fixed center. We denote its position by the coordinates $\Phi_0 = \varphi_0$ and $\Theta_0^* = \theta_0^*$. All gyres are located in a limited band of latitudes, excluding areas near the North and South Poles ($\Theta^* = \pm\frac{\pi}{2}$), where the Jacobian coefficients turn to zero and infinity, respectively. Furthermore, all vortices lie on the same side of the equator, so that $\sin \Theta^*$ (and the third term on the left-hand side of Eq. (17.39)) does not change signs in the fluid flow. Taking this into account, we make the replacement in trigonometric multipliers of Eq. (17.39),

$$\Theta^* = \Theta_0^*,$$

(17.40)

and assume that $\lambda = \cos^2 \Theta_0^*$ as well. Then, Eq. (17.39) has the form

$$\frac{D(X_t, X)}{D(\varphi, \theta^*)} + \frac{D(Y_t, Y)}{D(\varphi, \theta^*)} = (S_1^* - 2\Omega D_0^* \sin \Theta_0^*) R_0^{-2} \cos^2 \Theta_0^* = S_1^{**}.$$

(17.41)

Condition (17.40) can be called the "averaged latitude approximation," and Eq. (17.41) is the equation for the radial Cauchy invariant in the averaged latitude approximation.

There are two types of terms in Eq. (17.39). The first two terms in square brackets (including Jacobians) describe the vortex dynamics indirectly related to the effect of the Earth's rotation. The third term already contains Ω and therefore takes into account the effect of the

316 *Analytical Fluid Dynamics in Lagrangian Variables*

planet's rotation. The averaging procedure in this term seems to be the most natural. Since the spin vorticity varies monotonically with latitude, upon replacing it with the average value, we assume that the effect of rotation acts in the same way over the entire latitude band. This property of our approximation is similar to the f-plane approximation. But we remain within a spherical coordinate system where the latitude and longitude functions are unknown. Here, continuity equation (17.38) is used in the exact formulation. The approximation is applied only to the Cauchy integral equation. Averaging of the Jacobian coefficients means that the weights of the meridional and azimuthal components in the Cauchy integral will vary. But since the Jacobians enter Eqs. (17.38) and (17.41) as identical linear combinations (with different weight coefficients), we can assume that the solutions of the set of approximate equations will not differ much from the solutions of (17.38) and (17.39). In the mathematical sense, we replace the coefficients of the nonlinear equation, which are monotonic and sign-constant functions, with their averaged values. Obviously, the approximate solution will be the more accurate the narrower the band of latitudes inside the gyre and the closer the part of the flow to the gyre center.

Equations (17.38) and (17.41) coincide with the equations of two-dimensional dynamics of an ideal incompressible fluid in the Cartesian Lagrangian coordinates (see Eqs. (1.11) and (1.23)). The functions X and Y are similar to the horizontal and vertical coordinates of the trajectory of a fluid particle, respectively, and φ and θ^* refer to its Lagrangian variables.

Introducing the complex coordinates $W = X + iY$ and $\overline{W} = X - iY$ of the particle trajectory and the complex Lagrangian coordinates $= \varphi + i\theta^*$ and $\overline{\chi} = \varphi - i\theta^*$, we can write Eqs. (17.38) and (17.41) in the form

$$\frac{D(W,\overline{W})}{D(\chi,\overline{\chi})} = R_0^{-2} D_0^* \cos^2 \Theta_0^* = D_0^{**}(\chi,\overline{\chi}), \qquad (17.42)$$

$$\frac{D(W_t,\overline{W})}{D(\chi,\overline{\chi})} = \frac{i}{2} S_1^{**}(\chi,\overline{\chi}). \qquad (17.43)$$

The functions D_0^{**} and S_1^{**} are independent of time. The first one determines the dependence of the initial position of fluid particles in Lagrangian coordinates.

Vortices in a Rotating Fluid 317

The sign of the function does not change in the flow region because the coordinates of the particles and their Lagrangian labels match one to one. We assume for definiteness that $D_0^{**} \geq 0$. When writing Eq. (17.41), we took into account the fact that the time derivative of Eq. (17.42) is zero.

17.3.3. *Ptolemaic vortices on the sphere*

Equations (17.40) and (17.41) have an exact solution (see Chapter 7)

$$W = W_0(t) + G(\chi - \chi_0)e^{i\delta t} + F(\overline{\chi} - \overline{\chi}_0)e^{i\mu t}, \qquad (17.44)$$

where $W_0(t)$ is a complex function of time, G and F are analytical functions of the corresponding arguments, χ_0 is a complex constant, and δ and μ are real valued. It is convenient to assume that the function W_0 determines the position of the vortex center, which remains stationary in our consideration, and therefore W_0 is constant and equal to

$$W_0^* = \Phi_0 \cos^2 \Theta_0^* + i \sin \Theta_0^*. \qquad (17.45)$$

Suppose we have a complex constant in the form $\chi_0 = \varphi_0 + i\theta_0^*$, where φ_0, θ_0^* are the Lagrangian coordinates of the gyre center ($\varphi_0 = \Phi_0$, $\theta_0^* = \Theta_0^*$) and require $G(0) = F(0) = 0$, so that $W(\chi_0, \overline{\chi}_0, t) = W_0^*$.

In the X, Y plane, the motion of the fluid particle will be a superposition of two circular rotations. A circular motion relative to the point W_0^* with radius $|G|$ and frequency δ is superimposed on a circular rotation with radius $|F|$ and frequency μ. The particle trajectories for this solution are either epicycloids ($\delta\mu > 0$) or hypocycloids ($\delta\mu < 0$). In the Ptolemaic picture of the world, the planets move along such trajectories, so these flows were called Ptolemaic (Section 7.2).

Let us introduce the notation $\chi' = \chi - \chi_0$ and write the expression (17.44) as follows:

$$W - W_0^* = G(\chi')e^{i\delta t} + F(\overline{\chi'})e^{i\mu t}. \qquad (17.46)$$

For fixed χ', Eq. (17.46) gives a parametric representation of the trajectory of the fluid particle. The first term on the right-hand side of Eq. (17.46) describes the rotation along a circle of radius $|G|$. We will assume that $|G| > |F|$. The value of this radius is specified by

318   *Analytical Fluid Dynamics in Lagrangian Variables*

choosing the range of variation of χ'. Let this be a circle $|\chi'| \leq \alpha$. In the case $|G| > |\mu F/\delta|$, the trajectory of the particle will represent a shortened epicycloid or hypocycloid. In the case of equality $|G| = |\mu F/\delta|$, it will be an ordinary epicycloid or hypocycloid with cusps. Finally, in the case of the opposite inequality, the trajectory of the fluid particle will be an elongated epicycloid or hypocycloid having self-intersection points. If the ratio of frequencies μ and δ is an integer, then the trajectory will be a closed curve with $|(\mu/\delta) - 1|$ cusps. If the frequency ratio is rational, the trajectory will be closed after a finite number of revolutions. In the case of an irrational ratio between μ and δ, the trajectory will remain open.

All of these statements will qualitatively remain valid on the Earth's surface according to the formulas

$$\Phi = \Phi_0 + \cos^{-2}\Theta_0^* \mathrm{Re}[G(\chi')e^{i\delta t} + F(\overline{\chi}')e^{i\mu t}],$$
$$\Theta^* = \arcsin\{\sin\Theta_0^* + \mathrm{Im}[G(\chi')e^{i\delta t} + F(\overline{\chi}')e^{i\mu t}]\}. \tag{17.47}$$

In the Φ, Θ^* plane, trajectories of fluid particles are non-uniformly stretched and represent deformed epicycloids or hypocycloids. In this sense, we can speak about a qualitative analogy of fluid motion in the auxiliary plane of variables X and Y and in the physical plane of meridians Φ and parallels Θ^*. Equations (17.46) and (17.47) describe the superposition of two rotational motions. The first one is performed with a period $2\pi/\delta$, which corresponds to the rotation of the fluid particle along the great circle. The second one with period $2\pi/\mu$ corresponds to a perturbation of finite amplitude superimposed on the background flow. It is natural to attribute this perturbation to mesoscale vortices formed at the boundary of the more global vortices. The size of the mesoscale vortices is much smaller than the characteristic scale of the larger-scale vortices and their velocities are much higher. Therefore, it is natural to consider $\delta \ll \mu$, so that the trajectory of a fluid particle has a large number of cusps.

Equations (17.46) and (17.47) define the dynamics of the vortex flow corresponding to a circle of radius α in the plane of the Lagrangian variable χ'. Its value is determined depending on the position of the gyre at the Earth's surface. Obviously, α should not exceed the minimum distance from the center of the gyre Φ_0, Θ_0^* to the coastline. The shape of the vortex depends on two functions,

Vortices in a Rotating Fluid   319

G and F. This means that Eqs. (17.46) and (17.47) give a solution for a fairly wide class of initial conditions, including the shape of the vortical region and its inner velocity distribution at the initial time. The only constraint on the choice of functions G and F is the constant sign of the right-hand side of the continuity equation (17.42) inside the flow. Let us assume for definiteness that $D^{**} \geq 0$. In this case, the following condition must be fulfilled:

$$|G'|^2 - |F'|^2 \geq 0, \quad |\chi'| \leq \alpha. \tag{17.48}$$

The equality $|\chi'| = \alpha$ is allowed at the vortex edge.

Passing to specific examples, let us begin with the analysis of stationary gyres. The fluid motion is stationary if the Lagrangian quantities are invariant under time translation [2]. For a two-dimensional flow, this means that the coordinates of a fluid particle depend only on two variables $q, s + ut$, where q and s are Lagrangian variables and u is a constant velocity. The coordinate s varies along the streamline, which coincides with the trajectory of the fluid particle in a stationary flow, and the coordinate q varies from one streamline to another. In our case, these variables are conveniently chosen in the following form:

$$q = |\chi'|, \quad s = \arg\chi';$$

the function G is linear and the function F is power law,

$$G(\chi') = \chi' = q^{is}, \quad F(\overline{\chi'}) = \beta\overline{\chi'}^n = \beta q^n e^{-ins}, \tag{17.49}$$

where n is a positive integer and β is a constant with the dimension α^{1-n}. It follows from the condition (17.48) that $\beta \leq (n\alpha^{n-1})^{-1}$.

In the new variables, the expression (17.46) will be written as follows:

$$W - W_0^* = q \exp[i(s + \delta t)] + \beta q^n \exp\left[-in\left(s - \frac{\mu}{n}t\right)\right]. \tag{17.50}$$

Under the assumption $\mu = -n\delta$, the flow (17.50) is stationary since it depends only on two variables $q, s + \delta t$. The trajectories of fluid particles in the plane W and hence the stream functions are hypocycloids with $n + 1$ cusps. The fluid streamlines are deformed hypocycloids in the meridian-to-latitude Cartesian grid. The qualitative form of

320   *Analytical Fluid Dynamics in Lagrangian Variables*

streamlines for such vortices is simple to represent. They form a system of closed wavy lines around the gyre center with coordinates Φ_0 and Θ_0^*. The number of waves stowed along each of them is $n+1$.

In the case of an arbitrary ratio of frequencies δ and μ, the fluid motion is unsteady according to Eq. (17.50). The latter expression can be written as follows:

$$W - W_0^* = \{q \exp(i\Delta) + \beta q^n \exp(-in\Delta)\} \exp\left[i\left(\frac{\mu + n\delta}{n+1}\right)t\right],$$

$$\Delta = s + \frac{\delta - \mu}{n+1}t. \tag{17.51}$$

When $\mu = -n\delta$, the expression in braces describes a steady flow. The exponential multiplier outside the braces refers to the uniform rotation of the fluid. Then, the solutions (17.50) and (17.51) describe a hypocycloidal gyre rotating with angular frequency $(\mu + n\delta)/(n+1)$ without any shape variation.

For the considered flows, the stream function is introduced by the following formulas:

$$V_\Theta = R_0 \Theta_t = \frac{1}{\sin \Theta}\psi_\Phi, \quad V_\Phi = R_0 \Phi_t \sin \Theta = -\psi_\Theta.$$

The total differential for this function has the form

$$d\psi = \psi_\Phi d\Phi + \psi_\Theta d\Theta = \frac{R_0}{2i\lambda}\frac{D(W,\overline{W})}{D(\chi',t)}d\chi + c.c.$$

On the other hand, $d\psi = \psi_{\chi'}d\chi' + \psi_{\overline{\chi'}}d\overline{\chi'} = \psi_{\chi'}d\chi' + c.c.$ Therefore,

$$\psi_{\chi'} = \frac{R_0}{2i\lambda}\frac{D(W,\overline{W})}{D(\chi',t)},$$

so the stream function for the flows (17.46) is written in the form

$$\psi = -\frac{R_0}{\lambda}\{\delta|G|^2 + \mu|F|^2 + (\delta + \mu)\mathrm{Re}[G\overline{F}\exp i(\delta - \mu)t]\}. \tag{17.52}$$

For hypocycloidal gyres (17.49) in the case $\mu = -n\delta$, the stream function depends only on q, and $s + \delta t$ should be the case for steady flows. It is easy to show that the $\psi = $ const lines correspond to this curve by using the parametric representation of the hypocycloid. In any other case, expression (17.52) will correspond to unsteady flows.

Figure 40 shows the dynamics of the boundary of the gyre under the following conditions:

$$G = \chi', \quad F = -0.2\left(\frac{\overline{\chi'}}{\alpha}\right)^3 + 0.1\left(\frac{\overline{\chi'}}{\alpha}\right)^4, \quad |\chi'| \le \frac{\pi}{15},$$

$$\Phi_0 = \frac{\pi}{4}, \quad \Theta_0^* = \frac{\pi}{6}, \quad \mu = 6\delta.$$

The gyre corresponds to a circle of radius 12° on the Lagrangian plane χ'. The center of the gyre is chosen at a point corresponding to a 45° west longitude and a 30° north latitude. It is located inside the North Atlantic gyre, east of the Sargasso Sea. The sign of the angular frequency δ is positive, which corresponds to the clockwise periodic rotation of the gyre with period $2\pi/\delta$. The trajectories of the fluid particles are closed curves with five cusps. Figure 17.1 shows the dynamics of the gyre boundary at different moments of time corresponding to eight values of the parameter δt. In addition to rotational motion, the gyre deforms in a rather complex way as well. There are four cusps at the gyre boundary. They change their shape with time, so we can say that an unsteady azimuthal wave propagates along the gyre boundary.

It is interesting to compare the solution given by Eqs. (17.46) and (17.47) with the solution for Ptolemaic vortices in conventional

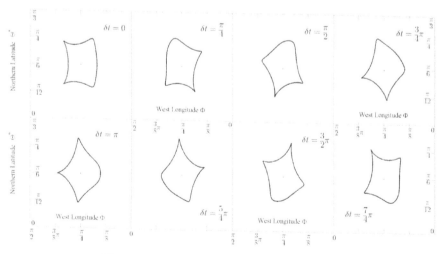

Fig. 17.1. Dynamics of the gyre boundary.

Analytical Fluid Dynamics in Lagrangian Variables

fluid dynamics. The velocity field of a two-dimensional plane vortex should decrease inversely proportional to the distance from its center at infinity. As a result, when matching the vortex motion with the external potential flow, the frequency μ is assumed to be zero and the trajectories of fluid particles inside the vortex are assumed to be circular (Section 7.2). In our case, $\mu \neq 0$, so fluid particles have more complex trajectories.

The Lagrangian invariants of the Ptolemaic flows on the sphere surfaces (17.46) and (17.47) are written as

$$D_0^* = \frac{R_0^2}{\cos^2 \Theta_0^*}(|G'|^2 - |F'|^2), \quad S_1^{**} = 2(\delta|G'|^2 - \mu|F'|^2),$$

$$S_1^* = \frac{R_0^2}{\cos^2 \Theta_0^*}[(\delta + \Omega)|G'|^2 - (\mu + \Omega)|F'|^2],$$

and the expression for (17.35) takes the form

$$\omega_R = \frac{S_1^*}{D_0^*} - (2\Omega - \cos^2 \Theta_0^* \mathrm{Re}W_t)\mathrm{Im}W.$$

In addition to the component depending only on the Lagrangian variables, it also contains non-stationary terms with frequencies $\delta, \mu, 2\delta, 2\mu, \mu + \delta$, and $\mu - \delta l$.

The vortex motions under study are induced by the pressure $P_s(\theta, \varphi)$ on the free surface. The pressure distribution from the set of Eqs. (17.18) and (17.19) can, in view of Eq. (17.34), be written as follows:

$$\Theta_{tt}\Theta_\theta + \sin^2 \Theta \Phi_{tt}\Phi_\theta + \cos \Theta \sin \Theta \left[2\Theta_t \Phi_t \Phi_\theta - \Theta_\theta \Phi_t^2 \right.$$

$$\left. + 2\Omega(\Theta_t \Phi_\theta - \Theta_\theta \Phi_t) - \Omega^2 \left(1 + \frac{2h}{R_0} \right) \right] = -\frac{1}{\rho R_0^2}(P_s)_\theta,$$

$$\tag{17.53}$$

$$\Theta_{tt}\Theta_\varphi + \sin^2 \Theta \Phi_{tt}\Phi_\varphi + \cos \Theta \sin \Theta \left[2\Theta_t \Phi_t \Phi_\varphi - \Theta_\varphi \Phi_t^2 \right.$$

$$\left. + 2\Omega(\Theta_t \Phi_\varphi - \Theta_\varphi \Phi_t) \right] = -\frac{1}{\rho R_0^2}(P_s)_\varphi. \tag{17.54}$$

To calculate the function P_s, it is necessary to find expressions for the variables Θ and Φ from Eq. (17.46) and then substitute them to the set of Eqs. (17.53) and (17.54) followed by integration. The left-hand sides of Eqs. (17.53) and (17.54) are extremely cumbersome expressions. It is impossible to calculate the integral explicitly even in the averaged latitude approximation. But in each case (for certain functions G and F), the pressure distribution on the free surface can be obtained by numerical integration. For unsteady vortex motions, the function P_s depends on time. Taking into account the fact that the choice of the functions G and F is rather arbitrary, we can conclude that Eqs. (17.53) and (17.54) define a wide class of surface pressure distributions. This gives us reason to assume the existence of natural Ptolemaic-type fluid flows in the ocean.

References

[1] Yakubovich, E.I. and Shrira, V.I. (2012). Non-steady columnar motions in rotating stratified Boussinesq fluids: Exact Lagrangian and Eulerian description, *J. Fluid Mech.*, 691, pp. 417–439.

[2] Bennett, A. (2006). *Lagrangian Fluid Dynamics* (Cambridge University Press, Cambridge).

[3] Abrashkin, A.A. (2019). Large Oceanic Gyres: Lagrangian Description. *J. Math. Fluid Mech.*, 21, 25.

[4] Constantin, A. and Johnson, R.S. (2017). Large gyres as a shallow-water asymptotic solution of Euler's equation in spherical coordinates, *Proc. R. Soc. A*, 473, 20170063.

Appendix 1

Speed and Acceleration in a Cylindrical Coordinate System

We will consider the fluid motion in the cylindrical coordinate system R, Φ, Z. At each point in space, we can construct a cylindrical basis consisting of vectors $\vec{e}_R$, $\vec{e}_\Phi$, and $\vec{e}_Z$ of unit length (basis vectors are directed tangentially to the coordinate line in the direction of increasing coordinates). The velocity (or acceleration) vector of a moving particle is expanded into the basis. It is possible to visualize one cylindrical base moving with the particle and simultaneously rotating. During this motion, the vector $\vec{e}_Z$ does not change its direction, while the vectors $\vec{e}_R$ and $\vec{e}_\Phi$ rotate around it. Since the length of these vectors does not change, $\dot{\vec{e}}_R \perp \vec{e}_R$ and $\dot{\vec{e}}_\Phi \perp \vec{e}_\Phi$ (the dot denotes differentiation in time). It follows that $\dot{\vec{e}}_R \parallel \vec{e}_\Phi$ and $\dot{\vec{e}}_\Phi \parallel \vec{e}_R$, i.e., $\dot{\vec{e}}_R = \alpha \vec{e}_\Phi$ and $\dot{\vec{e}}_\Phi = \beta \vec{e}_R$. To find the proportionality coefficients α and β, we decompose $\vec{e}_R$ and $\vec{e}_\Phi$ with respect to the Cartesian unit vectors $\vec{e}_X$ and $\vec{e}_Y$:

$$\vec{e}_R = \vec{e}_X \cos\Phi + \vec{e}_Y \sin\Phi, \quad \vec{e}_\Phi = -\vec{e}_X \sin\Phi + \vec{e}_Y \cos\Phi. \quad (A1.1)$$

In these equations, the angle Φ depends on time, and $\vec{e}_X$ and $\vec{e}_Y$ do not change with time. We differentiate Eq. (A1.1) in time:

$$\begin{aligned} \dot{\vec{e}}_R &= -\vec{e}_X \Phi_t \sin\Phi + \vec{e}_Y \Phi_t \cos\Phi, \\ \dot{\vec{e}}_\Phi &= -\vec{e}_X \Phi_t \cos\Phi - \vec{e}_Y \Phi_t \sin\Phi. \end{aligned} \quad (A1.2)$$

326   *Analytical Fluid Dynamics in Lagrangian Variables*

Comparing Eqs. (A1.1) and (A1.2), we obtain the desired formulas:

$$\dot{e}_R = \Phi_t \vec{e}_\Phi, \quad \dot{e}_\Phi = -\Phi_t \vec{e}_R. \tag{A1.3}$$

Now, let us start determining the velocity and acceleration vectors of a fluid particle. Its radius is the vector $\overrightarrow{R_*} = R\vec{e}_R + Z\vec{e}_Z$. By differentiating this equality in time and taking into account Eq. (A1.3), we decompose the velocity vector $\vec{V}$ by a cylindrical basis:

$$\vec{V} = \overrightarrow{R_{*t}} = R_t\vec{e}_R + R\Phi_t\vec{e}_\Phi + Z_t\vec{e}_Z. \tag{A1.4}$$

If we take the time derivative of this expression, we obtain a representation of the acceleration of a fluid particle in the cylindrical coordinate system:

$$\vec{a} = \left(R_{tt} - R\Phi_t{}^2\right)\vec{e}_R + \left(R\Phi_{tt} + 2R_t\Phi_t\right)\vec{e}_\Phi + Z_{tt}\vec{e}_Z. \tag{A1.5}$$

This equation was used in Section 1.6.1.

Appendix 2

Speed and Acceleration in a Spherical Coordinate System

Consider a fluid particle moving along some trajectory in the spherical coordinate system R, Θ, Φ. At each point on this trajectory, one can construct a different spherical basis from the vectors $\vec{e_R}, \vec{e_\Theta}$, and $\vec{e_\Phi}$. But we can assume that the same basis, which changes with time, moves along with the particle. Let us calculate the time derivatives of the basis vectors.

Since the lengths of these vectors are constant, the relations $\dot{\vec{e}}_R \perp \vec{e}_R$, $\dot{\vec{e}}_\Theta \perp \vec{e}_\Theta$, and $\dot{\vec{e}}_\Phi \perp \vec{e}_\Phi$ are valid, i.e., the vector of the time derivative of each unit vector lies in the plane of the other two. Let us use the results obtained earlier for the cylindrical basis ρ, Φ, Z (in the formulas of Appendix 1, we should make the following replacement: $R \rightarrow \rho$). First, we decompose the spherical vectors $\vec{e_R}$ and $\vec{e_\Theta}$ into cylindrical unit vectors $\vec{e_\rho}$ and $\vec{e_Z}$:

$$\vec{e_R} = \vec{e_\rho}\sin\Theta + \vec{e_Z}\cos\Theta; \quad \vec{e_\Theta} = \vec{e_\rho}\cos\Theta - \vec{e_Z}\sin\Theta. \qquad (A2.1)$$

On differentiating this expression by time, we obtain

$$\dot{\vec{e}}_R = \Theta_t(\vec{e_\rho}\cos\Theta - \vec{e_Z}\sin\Theta) + \dot{\vec{e}}_\rho\sin\Theta;$$

$$\dot{\vec{e}}_\Theta = -\Theta_t(\vec{e_\rho}\sin\Theta + \vec{e_Z}\cos\Theta) + \dot{\vec{e}}_\rho\cos\Theta,$$

or, with allowance for Eqs. (A1.3) and (A2.1),

$$\dot{\vec{e}}_R = \vec{e_\Theta}\Theta_t + \vec{e_\Phi}\Phi_t\sin\Theta; \quad \dot{\vec{e}}_\Theta = -\vec{e_R}\Theta_t + \vec{e_\Phi}\Phi_t\cos\Theta. \qquad (A2.2)$$

327

328   *Analytical Fluid Dynamics in Lagrangian Variables*

To determine the derivative of the azimuthal unit vector, we take the second of relation (A1.3) and express $\vec{e}_\rho$ through $\vec{e}_R$ and $\vec{e}_\Theta$. As a result, we find

$$\dot{\vec{e}}_\Phi = -\Phi_t(\vec{e}_R \sin\Theta + \vec{e}_\Theta \cos\Theta). \tag{A2.3}$$

Since the radius vector of a moving particle is $\vec{R} = R\vec{e}_R$, its speed has the form

$$\vec{V} = \vec{R}_t = R_t\vec{e}_R + R\Theta_t\vec{e}_\Theta + R\Phi_t\sin\Theta\vec{e}_\Phi, \tag{A2.4}$$

and the acceleration

$$\begin{aligned}
\vec{a} = {} & (R_{tt} - R\Theta_t{}^2 - R\Phi_t{}^2\sin^2\Theta)\vec{e}_R \\
& + (R\Theta_{tt} + 2R_t\Theta_t - R\Phi_t{}^2\sin\Theta\cos\Theta)\vec{e}_\Theta \\
& + [(R\Phi_{tt} + 2R_t\Phi_t)\sin\Theta + 2R\Theta_t\Phi_t\cos\Theta]\vec{e}_\Phi.
\end{aligned} \tag{A2.5}$$

Equation (A2.6) was used in Section 1.6.2.

Index

B

Brunt–Väisälä frequency, 305
bubble, 41

C

Cauchy invariants, 14, 16–17, 21, 63,
233, 250, 289
Cauchy vorticity formula, 18
centrifugal forces, 287
Chaplygin dipole vortex, 128
Chaplygin elliptical vortex, 128
complex Lagrangian coordinates,
55
continuity equation, 3, 9, 14, 23
contraflexure points, 226
Coriolis acceleration, 287
Coriolis force, 287
critical point, 29
cylindrical coordinates, 22
cylindrical variables, 30

D

Dirichlet problem, 67

E

edge waves, 97, 229
Ekman spiral, 295
epicycloidal waves, 138
Euler method, 6

Eulerian approach, 4
exact solution, 58

F

f-plane approximation, 289, 292, 297
flows inertial, 56
fluid cone, 38
fluid ellipse, 33

G

generalized Kirchhoff vortices, 156
generalized Ptolemaic, 170
Gerstner wave, 91, 107, 206
Grad–Shafranov equation, 167
Guyon waves, 107, 121

H

Hankel–Kelvin theorem, 20
Helmholtz theorems, 15, 18
Hill vortex, 167

J

Jacobi matrix, 10, 14, 18, 253
John method, 60

K

Kelvin waves, 148
Kida vortex, 128
Kirchhoff vortex, 59

329

L

Lagrange method, 6
Lagrange theorem, 27, 96
Lagrangian approach, 4
Lagrangian coordinates, 5–6
Lagrangian variables, 8–9

M

material variables, 5
matrix equations, 14
modified Lagrangian coordinates, 108, 275
momentum equation, 3, 9, 15, 23, 28, 54
multiscale method, 81, 121

N

Navier–Stokes equation, 247, 250, 292
nonlinear dispersion relation, 78
nonlinear Schrödinger, 82, 121

P

point vortex, 32
ptolemaic flow, 130, 206
ptolemaic vortices, 148

R

Rankine circular vortex, 57
relabeling symmetry, 21

rotating ellipsoids, 63
Routh–Hurwitz criterion, 214, 221
run-up patterns, 237

S

Saffman–Moore vortex, 128
spherical coordinates, 307, 309
spiral trajectories, 59
standing, 266
standing wave, 120, 266
Stokes drift, 77, 94, 116, 275
Stokes expansion, 71
Stokes wave, 71, 94, 107

V

viscosity, 247
vortex core, 59
vortex line, 15
vortex tube, 15
vorticity, 15–16
vorticity flux, 19
vorticity vector, 17, 19

W

wave breaking, 223
wave train, 281

www.ingramcontent.com/pod-product-compliance
Lightning Source LLC
Jackson TN
JSHW011605180225
79191JS00002B/16